WATER TREATMENT EXAM PREPARATION FOR OPERATORS

WATER EXAMS COMPANY

WATER EXAMS
CO

g.a.waterexams@gmail.com

jimmy.gamblue@gmail.com

Jimmy Abdalla
Water chemist

Water Exams Link: https://www.facebook.com/disrtibutreatment

LinkedIn:

https://www.linkedin.com/in/gamaleldin-abdalla-096252315/?skipRedire ct=true

g.a.waterexams@gmail.com

jimmy.gamblue@gmail.com

بِسْمِ اللَّهِ الرَّحْمَٰنِ الرَّحِيمِ
بِاسْمِكَ اللَّهُمَّ

DEDICATION

To my father, **Engineer Abdalla Moghazy**

Your guidance, strength, and example have shaped the person I am today. This book is dedicated to you with love, respect, and gratitude.

إهداء

إلى والدي، المهندس عبد الله مغازي

إليك أهدي هذا الكتاب، عرفانًا بفضلك، وامتنانًا لعطائك الذي لا يُقدَّر بثمن

لقد كنتَ المعلِّم الأول، والقدوة الصالحة، والسند الذي منحني القوة لأواصل طريقي. بتوجيهك، وحكمتك، وصبرك، وإيمانك بي، تشكَّلت شخصيتي، وتعلَّمت أن النجاح ثمرةُ الاجتهاد، وأن العلم رسالة، وأن خدمة الناس من أشرف الغايات.

وما وصلتُ إليه اليوم ما كان ليكون، بعد فضل الله، إلا بدعمك وتشجيعك وتضحياتك التي ستبقى محفورةً في قلبي ما حييت.

أسأل الله أن يجزيك عني خير الجزاء، وأن يبارك في عمرك وصحتك، وأن يجعل هذا العمل في ميزان حسناتك، كما كان دعمك لي دائمًا سببًا في كل خطوةٍ من خطوات نجاحي.

إليك أهدي هذا الكتاب بكل الحب والتقدير والاحترام، والعرفان بالجميل.

ابنك المحب
JIMMY

Gamaleldin Abdalla (Jimmy)

Water treatment chemist, CALIFORNIA, USA

AUTHOR'S NOTE

I wrote this book to help people who are preparing for the T1 and T2 water treatment exams and working toward a state license in water treatment operations. Water treatment is more than a technical field. It is a career that protects public health, supports communities, and gives people the chance to build a stable future. My goal is to make the basic ideas of water science, water chemistry, treatment processes, and exam questions easier to understand.

This book comes from my love of helping people through science. I believe knowledge should be shared in a simple and useful way, especially with young adults and new operators who want a better life and a good career for themselves. I hope this book helps you study with confidence, understand the important concepts, and take one step closer to your license and your future.

Disclaimer

This book is intended for educational and exam preparation purposes only. It is designed to help readers review basic water treatment concepts, water chemistry, treatment processes, and practice questions for T1 and T2 water treatment exams. Water treatment rules, regulations, exam requirements, and operating procedures may vary by state, agency, and local authority. Readers should always refer to official state licensing materials, current regulations, approved training resources, and guidance from the appropriate regulatory agencies.

This book does not replace official training, professional instruction, regulatory guidance, or the responsibility of a certified operator. Always follow current federal, state, and local requirements when operating or maintaining a public water system.

Jimmy Abdalla

Water treatment chemist

TABLE OF CONTENT

CHAPTER 1

WATER CHEMISTRY, WATER BASICS, WATER SOURCES, CONTAMINANTS, AND PUBLIC WATER SYSTEMS

Water Treatment

Water treatment usually refer to the core steps used to turn raw surface water or groundwater into safe drinking water: coagulation, flocculation, sedimentation, filtration, and disinfection.

Why Do We Treat Water?

- Remove particles and turbidity so water is clear and pleasant to use.
- Remove or inactivate pathogens, such as bacteria, viruses, and protozoa, that cause disease.
- Control taste, odor, color, and corrosivity to protect pipes and infrastructure.

Main Drinking Water Sources

1. Surface Water

Examples: rivers, lakes, reservoirs, and impoundments.

Surface water typically has higher turbidity, more pathogens, more organic matter, and greater vulnerability to runoff and spills. Because of this, it usually requires full conventional treatment, including coagulation, flocculation, sedimentation, filtration, and disinfection.

2. Groundwater

Examples: wells, springs, and infiltration galleries.

Groundwater is often lower in turbidity and pathogens due to natural filtration through soil and rock. However, it can have higher dissolved minerals, such as hardness, iron, and manganese, and it can be vulnerable to contamination from septic systems, agriculture, and industrial sites.

Note: Some systems use "groundwater under the direct influence of surface water"

(GWUDI), which is treated more like surface water because it can carry surface-water characteristics and pathogens.

What is the Water Table?

The water table is the top surface of the underground zone where soil and rock are completely filled, or saturated, with water.

Above the water table, the ground has both air and water in the pores. This is called the unsaturated zone.

At the water table and below, all the pores are filled with water. This saturated zone is where groundwater and aquifers are found.

Extra Exam-Style Points

The water table level can rise after heavy rain or recharge and fall during droughts or heavy pumping from wells.

Wells used for drinking water must extend below the water table so water can flow from the saturated zone into the well.

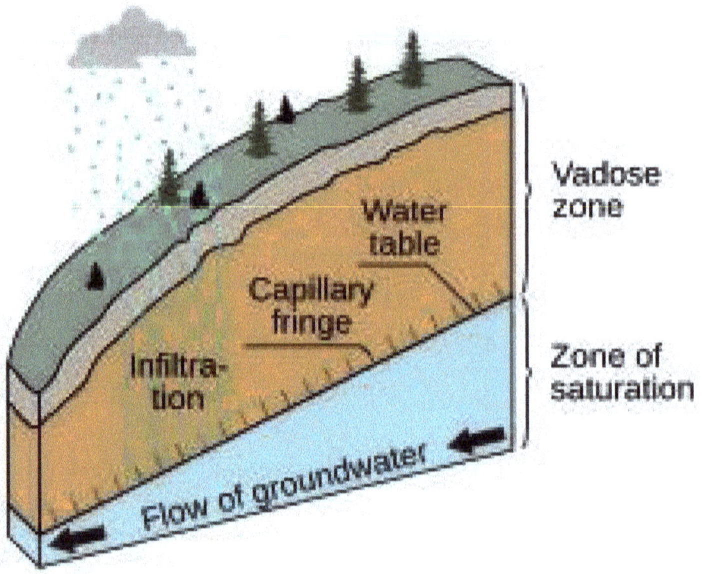

The Safe Drinking Water Act (SDWA)

The Safe Drinking Water Act (SDWA) is the main U.S. federal law that protects the quality of drinking water by setting enforceable standards for public water systems and their sources.

Core Purpose

Passed in 1974, the Act aims to protect public health by regulating public drinking water supplies, including both surface water and groundwater sources. It authorizes the U.S. Environmental Protection Agency (EPA) to set national drinking water standards and oversee states, utilities, and water suppliers that must comply with those standards.

WATER CONTAMINANTS

Water contaminants are commonly grouped into four main categories: physical, chemical, biological, and radiological.

Physical Contaminants

Physical contaminants affect the appearance or physical properties of water, such as cloudiness, color, and odor.

Examples include sediment, silt, sand, rust, organic debris, and algae that cause turbidity.

Chemical Contaminants

Chemical contaminants are elements or compounds that may be naturally occurring or man-made.

Inorganic contaminants: metals such as lead, arsenic, and copper, as well as nitrate, salts, and hardness minerals.

Organic contaminants: pesticides, solvents, petroleum products, disinfection byproducts such as THMs and HAAs, PFAS, pharmaceuticals, and personal care products.

Biological Microbial Contaminants

Biological, or microbial, contaminants are living organisms or microbes found in water.

Examples include bacteria, such as *E. coli*, viruses, protozoa, such as *Giardia* and Cryptosporidium, and parasites. These are the main cause of acute waterborne disease outbreaks.

Radiological Contaminants

Radiological contaminants are unstable elements that emit ionizing radiation. Examples include uranium, radium, cesium, and plutonium, which may occur naturally in some geologies or come from human activities.

Exam-Style Breakdown

Some references also summarize drinking water contaminants as:

- Inorganic compounds, such as arsenic, nitrate, and lead.
- Organic compounds, such as solvents, pesticides, and industrial chemicals.
- Disinfection byproducts, such as trihalomethanes and haloacetic acids.
- Radionuclides, such as uranium and radium.
- Infectious agents, such as bacteria, viruses, and protozoa.

WHAT DOES AN AQUIFER MEAN?

An aquifer is an underground layer of permeable rock, sand, or gravel that is saturated with water and can supply useful amounts of groundwater to wells or springs.

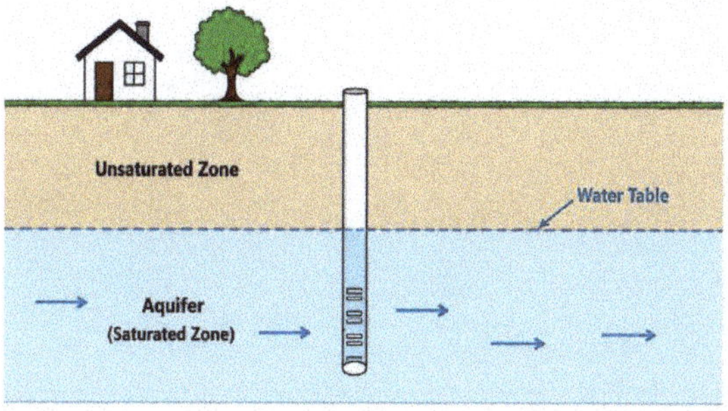

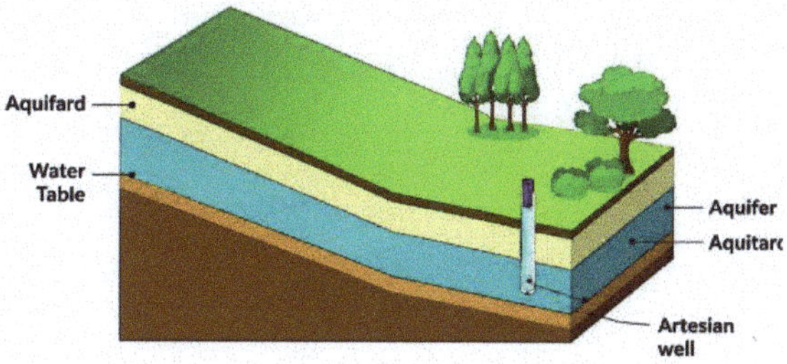

In simpler words, an aquifer is a water-bearing layer of rock or sediment below the ground.

It stores groundwater in the pore spaces between grains or in fractures, and it transmits that water so it can be pumped out.

It is not an open underground river or lake. It is more like a soaked sponge made of sand, gravel, or rock.

Sanitary Survey

A sanitary survey is an on-site inspection and overall review of a public water system's source, facilities, equipment, operation, and maintenance to evaluate its ability to provide safe drinking water.

Main Purpose

A sanitary survey helps identify existing or potential problems, also called deficiencies, that could contaminate the water or threaten system reliability and public health.

It also documents the system's technical, operational, and management capacity and determines whether technical assistance or improvements are needed. Typically, the survey covers eight core areas of a public water system:

- Source, such as wells, springs, and surface intakes
- Treatment processes and equipment
- Finished water storage, such as tanks and reservoirs
- Distribution system, including pipes, valves, and cross-connection control
- Pumps, pump facilities, and controls
- Monitoring, reporting, and data verification, including sampling and records
- System management and operations
- Operator compliance and overall system compliance with rules

ARTESIAN WELL

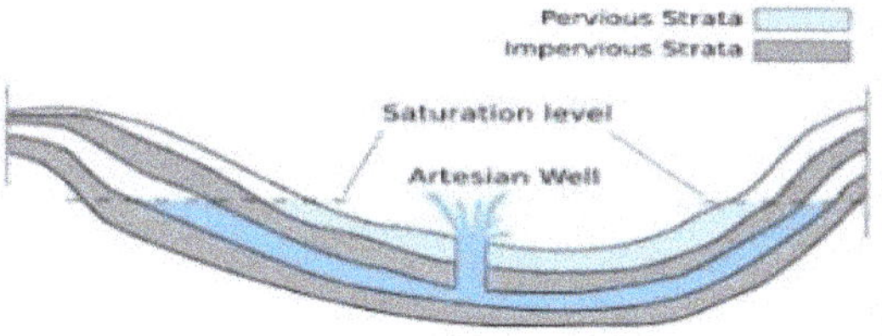

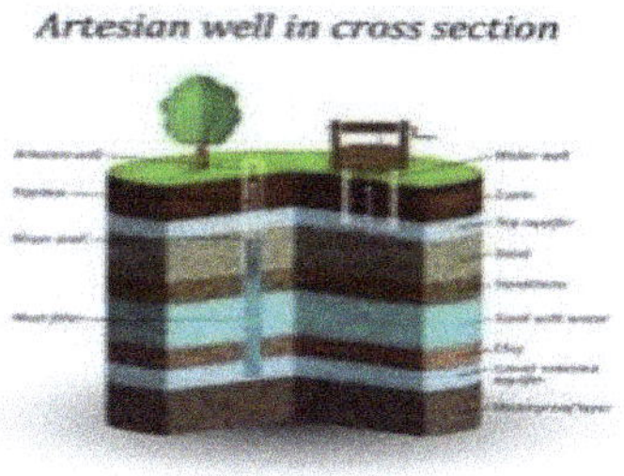

An artesian well is a groundwater well where water rises due to natural pressure in a confined aquifer, sometimes flowing out at the surface without any pump.

The well taps a confined aquifer, which is a permeable layer, such as sandstone, that holds water and is trapped between impermeable layers, such as clay or rock. Because the aquifer is under pressure, when you drill into it, the water level in the well rises to the "piezometric level." If that level is above the ground surface, the well flows on its own. This is called a flowing artesian well.

The water itself is normal groundwater. What makes it special is the pressure, not the chemistry.

A Public Water System

A public water system is a system that provides drinking water to the public through pipes or similar conveyances, serving at least 15 connections or 25 people for at least 60 days per year.

The word "public" refers to who is served, not who owns it. A public water system can be city-owned or privately owned.

Most systems in the U.S. are regulated under the Safe Drinking Water Act and must meet federal and state drinking water standards.

Main Types of Public Water Systems

Community Water System: Serves the same people year-round, such as cities, towns, and mobile home parks.

Non-Transient Non-Community Water System: Serves at least 25 of the same people for 6 or more months per year, such as schools or factories with their own wells.

Transient Non-Community Water System: Serves people who are there temporarily, such as gas stations or campgrounds with their own water system.

WHAT IS THE HYDROLOGIC CYCLE?

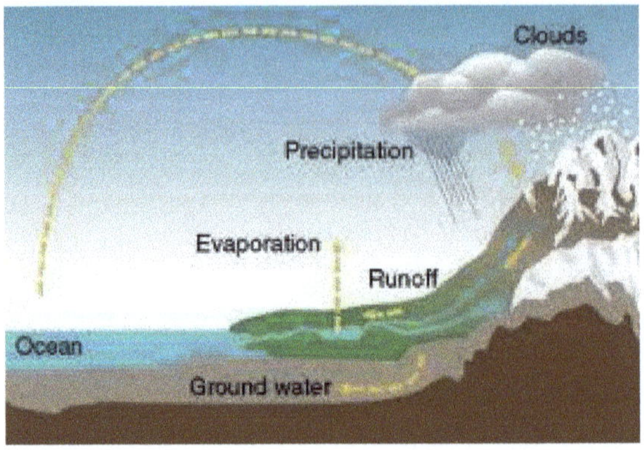

The hydrologic cycle, also called the water cycle, is the continuous movement of water on, above, and below Earth's surface. It is driven mainly by the Sun's energy. Water constantly circulates between oceans, the atmosphere, land, ice, surface water, and groundwater. However, the total amount of water on Earth stays essentially constant.

This cycle is crucial for climate, ecosystems, and human water supply, including rivers, aquifers, and soil moisture.

WHAT IS STRATIFICATION?

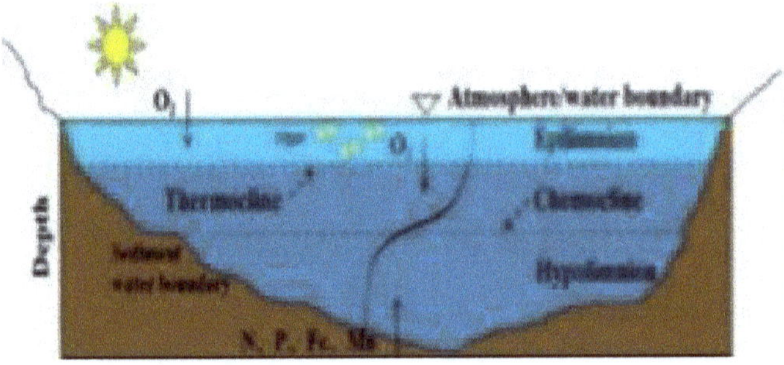

Stratification usually refers to thermal stratification in a reservoir or basin. This happens when water separates into layers of different temperature and density that do not mix well.

It is the formation of relatively stable layers in a body of water, with lighter and warmer water on top and denser, cooler water at the bottom.

It acts as a barrier to vertical mixing, so oxygen, heat, and nutrients do not move easily between the upper and lower layers.

In a stratified reservoir, the bottom layer, called the hypolimnion, can become low in dissolved oxygen and accumulate iron, manganese, ammonia, hydrogen sulfide, and other reduced compounds. These can cause taste, odor, and treatment issues when the water is pulled into the plant.

SURFACE WATER TREATMENT STEPS

Reservoir

The reservoir holds and stores the raw water before it goes into the treatment plant. **Screens**

Screens remove large objects from the water, such as sticks, leaves, fish, and trash.

Pre-Chlorination

Chlorine is added early to control algae and reduce taste and odor problems in the raw water. **Flash Mix**

Chemicals, such as coagulants, are quickly and thoroughly mixed with the water so they can start working.

Coagulation/Flocculation

Tiny particles in the water stick together to form larger clumps, called floc, that are easier to remove.

Sedimentation

The heavy floc settles to the bottom of a basin so it can be removed from the water.

Filtration

The water passes through filters, usually made of sand, gravel, and/or other media, to remove smaller particles that did not settle.

Disinfection

A disinfectant, often chlorine, is added to kill germs such as bacteria, viruses, and other disease-causing organisms.

Clear Well

Treated water is stored in the clear well, where the disinfectant gets more contact time and the water is held ready for high demand.

CONVENTIONAL SURFACE WATER TREATMENT PLANT

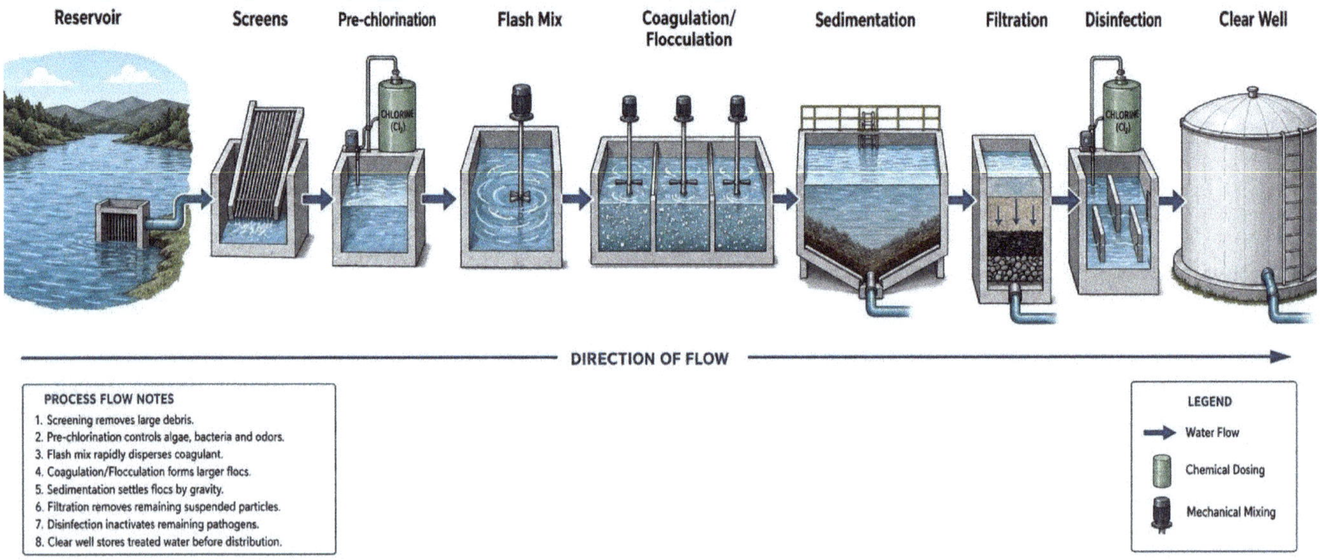

Trihalomethanes (THMs)

Trihalomethanes, or THMs, are disinfection byproducts that form when chlorine or other disinfectants react with natural organic matter in drinking water.

Chemically, a trihalomethane is a methane-type compound in which three of the four hydrogen atoms are replaced by halogens, such as chlorine or bromine. In drinking water treatment, "THMs" usually refers to a group of four compounds:

chloroform, bromodichloromethane, dibromochloromethane, and bromoform.

Why Do They Matter in Water Treatment?

THMs form mainly when chlorine is added to surface water that contains natural organic material, such as leaves, humic substances, and algae.

Long-term exposure to high THM levels has been associated with an increased risk of bladder and other cancers, as well as possible reproductive or developmental effects. Because of this, THMs are regulated in drinking water.

Key Control Ideas for EXAM Style Questions

Remove as much natural organic matter as possible before chlorination through good coagulation, flocculation, sedimentation, and filtration.

Optimize where and how much chlorine is added, or use alternative disinfectants or combined strategies, while still maintaining adequate disinfection.

Turbidity

Turbidity is how cloudy or murky water is because of tiny particles floating in it. It is a key indicator of water quality and filtration performance.

Turbidity is caused by suspended particles such as clay, silt, organic matter, algae, and microorganisms that scatter light.

It is usually measured in NTU, or Nephelometric Turbidity Units, with a Turbidimeter.

Why Does Turbidity Matter in Drinking Water?

High turbidity can "hide" bacteria and viruses, making disinfection with chlorine or UV less effective and increasing the risk of disease.

For systems using conventional or direct filtration, U.S. rules require turbidity to be ≤1.0 NTU at all times and ≤0.3 NTU in at least 95% of monthly samples. Many plants aim for <0.1 NTU.

Microorganisms and Pathogenic Microorganisms

All pathogens are microorganisms, but not all microorganisms are pathogens.

Simple Definitions

Microorganism: Any very small living thing, or virus, that usually requires a microscope to see. Examples include bacteria, viruses, fungi, protozoa, and algae.

Many microorganisms are harmless or even helpful.

Pathogenic organism, or pathogen: A microorganism that causes disease or harm to its host. It is defined by its ability to cause illness, also called pathogenicity or virulence.

Key Difference for Exam Wording

Microorganisms include both beneficial or harmless types and harmful types, such as normal flora, decomposers, and probiotics.

Pathogenic organisms are the harmful subset of microorganisms that invade a host, multiply, and damage cells or interfere with normal body functions, leading to disease.

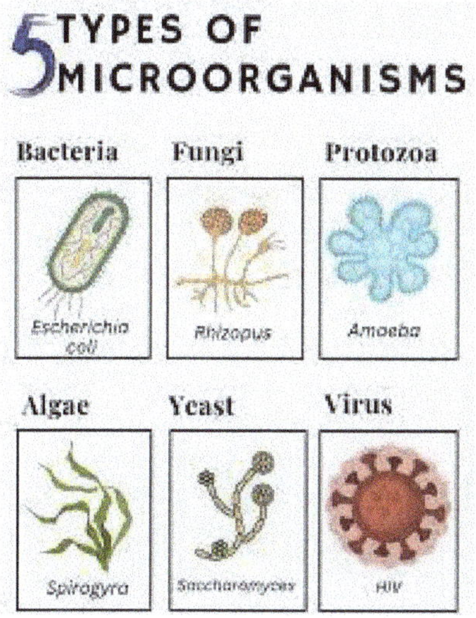

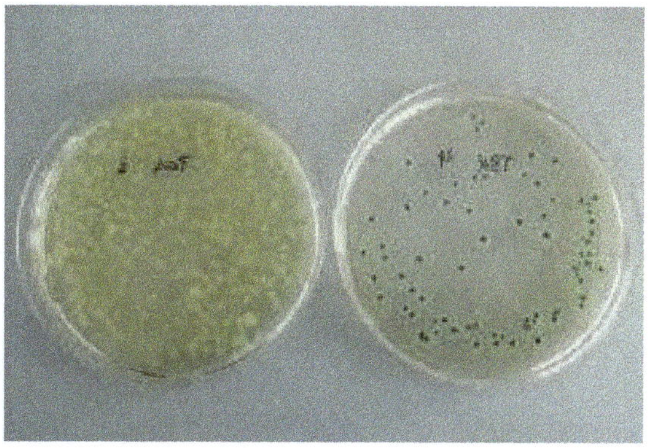

Basic Water Chemistry for Drinking Water

Basic water chemistry focuses on a few core ideas: pH, alkalinity, hardness, dissolved solids, and how they affect treatment, including coagulation, corrosion, scaling, and disinfection.

Key Properties

pH

- Measures how acidic or basic water is on a 0–14 scale. A pH of 7 is neutral.
- Reflects the balance between hydrogen ions and hydroxide ions.
- Strongly affects corrosion, scale formation, and disinfectant effectiveness in treatment plants and distribution systems.

Alkalinity

- Water's capacity to neutralize acids, mainly from bicarbonate, carbonate, and hydroxide ions.
- Acts as a buffer that resists rapid pH changes, which is critical for stable treatment and corrosion control.
- Often reported as mg/L as $CaCO_3$ and is closely tied to natural carbonate and hardness chemistry.

Hardness

Hardness is primarily the concentration of calcium and magnesium ions in water, usually expressed as mg/L as $CaCO_3$.

- High hardness promotes scale formation but can reduce the bioavailability of some toxic metals by precipitation.
- Softening processes, such as lime softening, ion exchange, and RO, are used where hardness causes operational or aesthetic problems.

Total Dissolved Solids (TDS)

- The sum of all dissolved inorganic salts and some organics in water.
- High TDS can affect taste, scaling, and suitability for industrial use, and often requires processes such as RO for reduction.

Organic Matter and DBP Precursors

- Natural organic matter reacts with chlorine and other oxidants, forming disinfection byproducts, or DBPs.
- Coagulation, filtration, and activated carbon are used to reduce organics before or along with disinfection.

Why These Matter for Treatment

- Coagulation, flocculation, and filtration performance depend on pH, alkalinity, and particle or organic chemistry.
- Disinfection effectiveness, including chlorine, ozone, and UV, depends on pH and the presence of organic and inorganic demand.
- Corrosion and scale control in distribution systems rely on balancing pH, alkalinity, and hardness to protect pipes while minimizing deposits.

SOME PROBLEMS IN THE GROUNDWATER TREATMENT PROCESS

Q: How can you solve bacteria contamination in groundwater?

Find and fix the contamination source, then disinfect the well through shock chlorination, continuous chlorination, or UV treatment. Protect the well from future entry of bacteria.

Q: How can you solve sulfide, or hydrogen sulfide, odors in well water?

Use aeration and/or oxidation, such as chlorine, peroxide, or ozone, followed by filtration to remove hydrogen sulfide gas and its odor.

Q: How can you solve hard groundwater?

Install an ion-exchange water softener to remove calcium and magnesium, and sometimes some iron, and replace them with sodium or potassium.

Q: How can you solve iron or manganese in groundwater?

Oxidize dissolved iron or manganese through aeration, chlorine, permanganate, or another oxidant, and then filter it out. Ion exchange or softening designed for iron and manganese removal may also be used.

Q: How can you solve high TDS in groundwater?

Use processes that remove dissolved salts, such as reverse osmosis, or RO.

Q: How can you solve corrosive well water?

Raise the pH and alkalinity with a neutralizing filter, such as calcium carbonate or magnesium oxide, or use a chemical feed, such as soda ash. Reduce aggressive metals if needed.

Q: How can you solve high nitrate in groundwater?

Use ion exchange with nitrate-selective anion resin or reverse osmosis. Also, address the pollution source, such as fertilizer or septic contamination, so nitrates stop entering the aquifer.

CHAPTER 1 EXAM QUESTIONS

1. Which of the following is an example of a surface-water source?

A. Deep confined aquifer

B. Infiltration gallery

C. River

D. Artesian well

Answer: C. River

2. Compared to groundwater, surface water usually has:

A. Lower turbidity and fewer pathogens

B. Higher turbidity and more pathogens

C. Higher hardness and more dissolved minerals

D. Lower organic matter

Answer: B.

3. Which of the following is an example of a groundwater source?

A. Lake

B. Reservoir

C. Spring

D. Estuary

Answer: C.

4. Groundwater typically has which characteristic compared to surface water?

A. More turbidity and more pathogens

B. Less turbidity and fewer pathogens

C. Lower dissolved minerals

D. Always needs full conventional treatment

Answer: B

5. What is the main concern with groundwater regarding dissolved constituents?

A. High algae growth

B. High organic color

C. High dissolved minerals, such as hardness, iron, and manganese

Answer: C.

6. What is the water table?

A. The top of a reservoir

B. The level where soil and rock first become saturated with water

C. The bottom of a lake

D. The interface between two rivers

Answer: B.

7. Above the water table, the soil pores contain:

A. Only water

B. Only air

C. Both air and water, also called the unsaturated zone

D. Oil and gas only

Answer: C.

8. An aquifer is best described as:

A. An open underground river

B. A water-bearing layer of rock or sediment that can supply useful groundwater

C. A storage tank in a treatment plant

D. A man-made lined basin for sludge

Answer: B.

9. What is "groundwater under the direct influence of surface water" (GWUDI)?

A. Groundwater that never changes temperature

B. Groundwater that is completely protected from contamination

C. A groundwater source that can carry surface-water characteristics and pathogens

D. Groundwater used only for irrigation

Answer: C.

10. What is the main purpose of a sanitary survey of a public water system?

A. To determine water rates

B. To inspect and evaluate the system's ability to provide safe drinking water

C. To design a new treatment plant

D. To test fire flow in hydrants only

Answer: B.

11. Which of the following is NOT one of the typical areas reviewed during a sanitary survey?

A. Source and treatment processes

B. Finished water storage

C. Distribution system and pumps

D. Customer billing procedures

Answer: D.

12. A public water system is defined, for drinking water rules, mainly by:

A. Ownership by the city

B. Serving at least 15 service connections or 25 people for at least 60 days per year

C. Having at least one storage tank

D. Using only surface-water sources

Answer: B.

13. Bacteria contamination in a well is usually controlled by:

A. Aeration only

B. Softening only

C. Finding and fixing the source and disinfecting the well through shock chlorination, continuous chlorination, or UV treatment

D. Adding fluoride

Answer: C

14. The "rotten egg" odor in well water is usually caused by:

A. Iron

B. Manganese

C. Hydrogen sulfide

D. Fluoride

Answer: C.

15. A common way to treat hydrogen sulfide odor in well water is to:

A. Add fluoride

B. Use aeration and/or oxidation followed by filtration

C. Remove hardness with ion exchange only

D. Add phosphate corrosion inhibitor

Answer: B.

16. The Safe Drinking Water Act (SDWA) is mainly intended to:

A. Regulate wastewater discharges to rivers

B. Protect drinking water quality by setting enforceable standards for public water systems and their sources

C. Regulate bottled water only

D. Set building plumbing codes

Answer: B

17. Under the SDWA, the main federal agency responsible for setting national drinking water standards is:

A. OSHA

B. CDC

C. EPA

D. USGS

Answer: C

18. The SDWA applies primarily to:

A. Private domestic wells serving one home

B. All public water systems that provide water for human consumption

C. Only surface-water treatment plants

D. Only groundwater systems in large cities

Answer: B

19. Which type of regulation under the SDWA sets legally enforceable limits on contaminants?

A. Primary drinking water standards

B. Secondary drinking water standards

C. Recommended operational guidelines

D. Emergency response plans

Answer: A

20. Secondary drinking water standards mainly address:

A. Acute health effects from pathogens

B. Aesthetic issues such as taste, odor, and color

C. Chlorine residual requirements

D. Corrosion control only

Answer: B

21. Drinking water contaminants are commonly grouped into which four broad categories?

A. Physical, chemical, biological, and radiological

B. Solid, liquid, gas, and plasma

C. Organic, inorganic, metallic, and plastic

D. Soluble, insoluble, volatile, and non-volatile

Answer: A

22. Physical contaminants in water mainly affect:

A. Pathogenicity of microbes

B. Appearance or physical properties, such as cloudiness, color, and odor

C. Radioactivity of water

D. Disinfection byproduct formation only

Answer: B

23. Which is the best example of a physical contaminant?

A. Nitrate

B. Sodium chloride

C. Sand and silt causing turbidity

D. Lead

Answer: c

24. Chemical contaminants include:

A. Only metals

B. Elements or compounds, naturally occurring or man-made

C. Only pesticides and solvents

D. Only synthetic organic chemicals

Answer: B

25. Which of the following is an example of an inorganic chemical contaminant?

A. Benzene

B. Atrazine

C. Nitrate

D. Trihalomethanes

Answer: C

26. Which is a common organic chemical contaminant in drinking water?

A. Calcium hardness

B. Arsenic

C. Pesticides and solvents

D. Radium

Answer: C

27. Radiological contaminants are best described as:

A. Colored dyes in the water

B. Unstable elements that emit ionizing radiation

C. Any dissolved salts

D. Only radioactive hydrogen, or tritium

Answer: B

28. Which statement is correct about microorganisms and pathogens?

A. All microorganisms are pathogens

B. All pathogens are microorganisms, and all microorganisms are harmless

C. All pathogens are microorganisms, but not all microorganisms are pathogens

D. Pathogens are always viruses, never bacteria

Answer: C

29. Pathogenic microorganisms are defined primarily by their:

A. Ability to grow in surface water

B. Ability to cause illness, or pathogenicity/virulence, in a host

C. Resistance to chlorine

D. Ability to form spores

Answer: B

30. Which group contains only examples of biological, or microbial, contaminants?

A. Sand, silt, rust, and algae

B. *E. coli*, *Giardia*, *Cryptosporidium*, and viruses

C. Lead, arsenic, and nitrate

D. THMs, HAAs, and PFAS

Answer: B

31. Microbial contaminants are especially important because they:

A. Only cause long-term chronic health effects

B. Are the main cause of acute waterborne disease outbreaks

C. Never survive in chlorinated systems

D. Only occur in groundwater

Answer: B

32. Which of the following is most closely associated with hardness in groundwater?

A. Calcium and magnesium

B. Benzene and toluene

C. Radium and uranium

D. PFAS compounds

Answer: A

33. Disinfection byproducts, or DBPs, such as Trihalomethanes (THMs) and Haloacetic acids (HAAs), generally form when:

A. Fluoride reacts with iron

B. Chlorine or other disinfectants react with natural organic matter in water

C. Lime is added for softening

D. Alum reacts with alkalinity

Answer: B

34. THMs are of concern because long-term exposure to high levels has been associated with:

A. Acute gastrointestinal illness only

B. Increased risk of bladder and other cancers and possible reproductive effects

C. Only skin irritation

D. Only tooth discoloration

Answer: B

35. Which source water condition tends to increase THM formation potential?

A. Very low organic carbon

B. High natural organic matter from leaves, humic substances, and algae

C. Very low temperature

D. Very low pH

Answer: B

36. Turbidity is best described as a measure of:

A. Dissolved solids concentration

B. Water hardness

C. Cloudiness caused by suspended particles

D. Radioactivity level

Answer: C

37. For filtered surface-water plants under the current rules discussed in your notes, turbidity must generally be:

A. Below 5.0 NTU at all times

B. Below 1.0 NTU at all times and $\leq$0.3 NTU in at least 95% of monthly samples

C. Exactly 0.1 NTU at all times

D. Below 10 NTU in 90% of samples

Answer: B

38. Low finished-water turbidity after filtration is important mainly because:

A. It proves the water is hard

B. It indicates good removal of particles and pathogens and improves disinfection effectiveness

C. It increases THM formation

D. It removes all dissolved inorganics

Answer: B

39. Compared to surface water, groundwater is more likely to:

A. Have higher turbidity and more pathogens

B. Be clearer with fewer pathogens but have higher dissolved minerals

C. Always require full conventional treatment

D. Contain more algae and color

Answer: B

40. "Groundwater Under the Direct Influence of Surface Water" (GWUDI) means:

A. Groundwater is completely isolated from surface contamination

B. Groundwater that is used only for irrigation

C. Groundwater that shows surface-water characteristics and may carry surface pathogens

D. Groundwater was found deeper than 1,000 feet

Answer: C

41. When finished water pH is consistently around 5.5, the main concern for the distribution system is:

A. Scale formation in pipes

B. Corrosion of metal pipes and fixtures

C. Reduced chlorine effectiveness only

D. Increased hardness deposits

Answer: B

42. Alkalinity in drinking water mainly:

A. Measures hardness only

B. Buffers pH against rapid changes

C. Measures chloride concentration

D. Has no impact on corrosion control

Answer: B

43. Water hardness is primarily caused by:

A. Sodium and potassium

B. Calcium and magnesium

C. Nitrate and nitrite

D. Sulfate and chloride

Answer: B

44. A common way to reduce hardness in a small groundwater system is:

A. Add fluoride

B. Aeration only

C. An ion-exchange softener using sodium or potassium

D. Chlorination only

Answer: C

45. Trihalomethanes (THMs) in treated water form mainly when:

A. Chlorine reacts with dissolved minerals

B. Chlorine reacts with natural organic matter

C. Chlorine reacts with fluoride

D. Chlorine reacts with dissolved oxygen

Answer: B

46. Which operational change will generally decrease THM formation?

A. Increasing natural organic matter in the raw water

B. Increasing detention time in warm storage tanks

C. Removing organic matter before chlorination, such as through better coagulation and filtration

D. Raising chlorine dose at the end of the system

Answer: C

47. High nitrate in a groundwater well is usually associated with:

A. Industrial metal discharges

B. Fertilizer and septic system contamination

C. Natural hardness from limestone

D. High algae growth in reservoirs

Answer: B

48. An effective treatment for high nitrate in a drinking water well is:

A. pH adjustment only

B. Ion exchange using nitrate-selective anion resin or reverse osmosis

C. Simple aeration tower

D. Sand filtration only

Answer: B

49. A typical treatment sequence for iron and manganese in groundwater is:

A. Fluoridation, then chlorination

B. Aeration or oxidation followed by filtration

C. UV disinfection only

D. Softening with no filtration

Answer: B

50. A small system has very low alkalinity and low-pH groundwater causing copper pinhole leaks. A good corrective action is:

A. Add more chlorine to kill corrosion bacteria

B. Reduce hardness with softening

C. Raise pH and alkalinity with soda ash or a neutralizing filter

D. Add coagulant at the well

Answer: C

CHAPTER 2

WATER FUNDAMENTALS, RESERVOIRS, AEROBIC AND ANAEROBIC CONDITIONS, ALGAE, AND ION EXCHANGERS

Acre-Foot

An acre-foot is the volume of water that would cover 1 acre of land to a depth of 1 foot.

It equals 43,560 cubic feet, or about 325,829 gallons.

Operators see acre-feet used for:

- Reservoir storage capacity
- Surface water rights and allocations
- Annual production or demand planning

Aeration

Aeration is the process of adding air, or oxygen, to water.

Common purposes include:

- Stripping gases, such as H_2S and CO_2, and reducing odors
- Oxidizing iron and manganese so they can be filtered
- Increasing dissolved oxygen for biological processes

In small groundwater systems, packed-tower aeration or cascade aerators are common.

Aerobic vs. Anaerobic

Aerobics:

- Oxygen is present as dissolved O_2.
- Supports aerobic bacteria that help break down organics in a controlled way.
- Typical of well-aerated lakes, aerated basins, and most distribution systems.

Anaerobic:

- Oxygen is not present.
- Other electron acceptors, such as nitrate or sulfate, are used by microorganisms.
- Can produce H_2S, or rotten egg odor, and methane, and can cause taste, odor, and corrosion problems in reservoirs, sludge, or dead-end zones.
- For T2, recognize that aeration can move water from anaerobic or low-oxygen conditions back toward aerobic conditions.

Algae

Algae are microscopic plants that contain chlorophyll and live in water. They may be floating or suspended, called planktonic, or attached to surfaces, called periphyton or biofilms.

They produce oxygen during the day through photosynthesis and use oxygen at night through respiration. Impacts on treatment include:

- Taste and odor problems
- Increased filter clogging and backwash frequency
- Increased organic matter, which can increase DBPs, such as THMs and HAA5, after chlorination

Algal Bloom

An algal bloom is a sudden, massive growth of algae in a lake or reservoir. It is often caused by:

- Warm temperatures
- Sunlight
- High nutrients, such as nitrogen and phosphorus Operational concerns include:
- Strong tastes and odors
- Slime and filter fouling
- Potential toxins from some cyanobacteria, or blue-green "algae" Control options include:
- Watershed nutrient control
- Mixing or aeration in reservoirs
- Careful use of algaecides, such as copper sulfate, following regulations

Coliform Bacteria

Coliform bacteria are a group of bacteria found in the intestines of warm-blooded animals, as well as in soil, plants, air, and water.

They do not always cause disease themselves, but they are used as an indicator that the water may contain pathogenic or disease-causing organisms.

In drinking water regulations:

- Total coliform presence triggers follow-up sampling and investigation.
- Fecal indicators, such as E. coli, are more directly linked to sewage or animal waste.

Colloids

Colloids are very small, finely divided particles that:

- Stay dispersed for a long time
- Do not settle easily by gravity
- Often carry an electrical charge, usually negative

Examples include very fine clays, organic color, and some metal complexes. They matter because they pass through plain sedimentation. They must be destabilized by coagulants, such as alum or ferric salts, so they can form larger floc.

THERMAL STRATIFICATION LAYERS

In a stratified lake or reservoir, especially in summer, water separates into layers by temperature.

- Epilimnion

The epilimnion is the upper layer of a stratified lake. It is warm, well mixed by wind, usually higher in dissolved oxygen, and often where algae grow because of sunlight and nutrients.

- Metalimnion

The metalimnion is the middle layer, also called the thermocline. The temperature drops rapidly with depth, and this layer acts as a barrier between warm upper water and cold lower water.

- Hypolimnion

The hypolimnion is the bottom layer in a stratified lake. It is the coldest layer and is often low in dissolved oxygen because it is isolated from the surface while organic matter decomposes.

It can become anaerobic, releasing iron, manganese, and hydrogen sulfide from sediments.

Operators care because intake depth, whether in the epilimnion or hypolimnion, affects temperature, dissolved oxygen, iron, manganese, taste, odor, and treatment needs.

ORGANIC SUBSTANCES

Organic substances are materials that come from plant or animal sources and contain carbon.

Examples in water include:

- Natural organic matter, such as leaves, decaying plants, and algae
- Humic and fulvic acids, which cause color
- Some pesticides, solvents, and other man-made organics

Organics are important because they use up disinfectant, creating chlorine demand.

They can also be THM precursors and can cause taste, odor, and color problems.

OXIDATION

Oxidation is the addition of oxygen to a substance, the removal of hydrogen, or the removal of electrons from a compound or element.

In practical water treatment, oxidation changes contaminants into a different form, often one that is easier to remove.

Examples include:

- $Fe^{2+} \rightarrow Fe^{3+}$, oxidized iron that can form particles and be filtered
- $Mn^{2+} \rightarrow$ higher-valence manganese, which is also more filterable
- Oxidizing hydrogen sulfide, or H_2S, to reduce odor

Common oxidants include chlorine, permanganate, ozone, and oxygen through aeration.

PH

pH is a measure of how acidic or basic, or alkaline, a liquid is.

- pH < 7 = acidic
- pH = 7 = neutral
- pH > 7 = basic, or alkaline pH controls coagulation performance because each coagulant has a best pH range. It also affects disinfection effectiveness because chlorine is more effective at lower pH.

pH also affects corrosion and scale in pipes. Very low pH is corrosive, while high pH with hardness can cause scaling. Operators constantly monitor and adjust pH to keep treatment and distribution stable.

THM PRECURSORS

THM precursors are natural organic compounds in water that can react with chlorine to form trihalomethanes, or THMs, which are disinfection byproducts. They come from decaying leaves, algae, other organic matter in surface water, and color-causing humic substances.

They are important because more organic matter, chlorine, time, and temperature can lead to more THMs. Regulations limit THMs in finished water, so operators must control precursors.

ALGAL BLOOMS AND WATER QUALITY

An algal bloom is a sudden, heavy growth of algae in a lake or reservoir. It creates several water quality and treatment problems.

Algal blooms can cause taste and odor problems. They can also lead to shorter filter runs because algae cells and slime plug filter media quickly. Filters reach terminal head loss faster, so more frequent backwashing is needed, increasing water and energy use.

During the day, algae use carbon dioxide for photosynthesis. Removing CO_2 tends to raise pH, which can affect coagulation, disinfection, and corrosion control. At night, algae and bacteria use oxygen through respiration. When a bloom dies and decomposes, microorganisms consume even more oxygen. This can drop dissolved oxygen and create low-oxygen or anaerobic conditions in the hypolimnion.

Algae also add a lot of organic matter to the water. This increases:

- Coagulant demand
- Filter loading
- THM precursors, because there are more organics to react with chlorine affect raw water quality throughout the year.

ANAEROBIC CONDITIONS AND PROBLEMS

- When dissolved oxygen is used up, water can become anaerobic, meaning no O_2 is present. This often happens in the hypolimnion and bottom sediments where organic matter is decomposing.

- One key problem in anaerobic conditions is sulfate reduction and H_2S odor.

Certain bacteria use sulfate instead of oxygen and convert it to sulfide. Sulfide combines with hydrogen to form hydrogen sulfide, or H_2S, which smells like rotten eggs. This causes taste and odor complaints and can increase corrosion.

- Another issue is iron and manganese. In anaerobic conditions at the reservoir bottom, iron and manganese become more soluble and enter the water from sediments.

When this water is later aerated or chlorinated, iron and manganese are oxidized and form particles that cause:

- Iron: red or brown stains on sinks, fixtures, and laundry
- Manganese: black stains on sinks and clothing

They also cause dirty-water complaints and can clog filters and mains.

- Algal bloom → more organics, pH up, dissolved oxygen swings, taste and odor, and shorter filter runs.
- Stratification → warm, oxygenated epilimnion on top and cold, low-dissolved-oxygen hypolimnion on the bottom.
- Anaerobic hypolimnion → H_2S, rotten egg odor, soluble iron and manganese, which later oxidize and stain.

COPPER SULFATE FOR ALGAE CONTROL

Copper sulfate is a common algaecide, but its effectiveness and safe dose depend on several water-quality factors.

Operators add copper sulfate to reservoirs or basins to kill or control algae growth. The actual dose that works without overdosing or wasting chemicals depends on:

- Alkalinity -PH - Suspended solids
- Organic matter
- Temperature
- Algae species

ALKALINITY

Alkalinity affects how available and effective copper is in the water.

In low-alkalinity water, usually 0–50 mg/L, copper sulfate is more effective at lower doses because more "free" copper ions are available to act on algae. When alkalinity is 50–150 mg/L, copper becomes less available, or more bound. A higher dose may be needed to get the same algaecide effect.

When alkalinity is greater than 150 mg/L, copper sulfate treatment is generally not recommended because high alkalinity greatly reduces effectiveness and increases the risk of using copper.

Simple memory: Low alkalinity = low dose works. High alkalinity = copper is less effective and may not be worth using.

PH: LOWER IS BETTER

Lower pH, meaning more acidic but still in a safe range, makes copper more soluble and effective against algae.

Higher pH tends to reduce copper's effectiveness and can increase precipitation of copper compounds.

For T2 level, think:

- Lower pH → more effective copper → possibly lower dose
- High pH, such as above 9 → copper treatment is much less effective

SUSPENDED MATTER: LOWER IS BETTER

Suspended solids, such as silt, clay, and turbidity, can bind or absorb copper, making less of it available to attack algae.

Clearer water with low suspended matter provides better algaecide efficiency.

ORGANIC MATTER: LOWER IS BETTER

Natural organic matter can also complex or bind copper, reducing its active form. High organic loading, such as decaying plants and leaves, means more copper may be required for the same effect. There is also more risk of DBP formation later when that organic matter is chlorinated.

For exam memory: High organics = more copper is "used up," leaving less available to kill algae.

Temperature: Higher is Better

Warmer water, which is typical in summer, improves algae metabolism and copper's effectiveness.

Algae do not respond well to treatment when water is below about 50°F. Copper sulfate is mainly useful in warmer seasons, not cold winter water. Key idea: Cold water + copper = poor response.

CHAPTER 2 REVISION

1. One acre-foot of water is equal to:

A. 43,560 gallons

B. 325,829 gallons

C. 1,000,000 gallons

D. 10,000 gallons

Answer: B

2. One acre-foot of water is:

A. The volume that covers one acre to a depth of one foot

B. The volume that covers one square foot to a depth of one acre

C. One million cubic feet

D. One gallon per minute for one day

Answer: A

3. Which process is defined as the addition of air to water?

A. Filtration

B. Aeration

C. Sedimentation

D. Chlorination

Answer: B

4. A condition in which dissolved oxygen is present in water is called:

A. Anaerobic

B. Anoxic

C. Aerobic

D. Colloidal

Answer: C

5. A condition in which dissolved molecular oxygen is not present in water is:

A. Aerobic

B. Anaerobic

C. Epilimnion

D. Oxidized

Answer: B

6. Algae are best described as:

A. Microscopic animals that live in sediments

B. Microscopic plants that contain chlorophyll and live in water

C. Bacteria that cause disease

D. Only organisms that grow in groundwater

Answer: B

7. What do algae do during the day?

A. Use oxygen and produce carbon dioxide

B. Produce oxygen and use carbon dioxide

C. Produce hydrogen sulfide

D. Do nothing

Answer: B

8. At night, algae primarily:

A. Produce oxygen

B. Stop all biological activity

C. Use oxygen, or respire

D. Turn into bacteria

Answer: C

9. An algal bloom is:

A. The death of all algae in a lake

B. A sudden and massive growth of algae in a lake or reservoir

C. Slow, steady algae growth over many years

D. Algae attached only to rocks

Answer: B

10. An anion is:

A. A positively charged ion

B. A negatively charged ion

C. A neutral particle

D. An ion with no electrons

Answer: B

11. A cation is:

A. A positively charged ion

B. A negatively charged ion

C. A neutral molecule

D. A dissolved gas

Answer: A

12. Coliform bacteria are important in drinking water because they:

A. Always cause disease

B. Indicate the possible presence of pathogenic organisms

C. Remove turbidity

D. Improve taste and odor

Answer: B

13. Coliform bacteria are commonly found in:

A. Only human intestines

B. Only industrial wastewater

C. Intestines of warm-blooded animals, as well as plants, soil, air, and water

D. Only deep groundwater

Answer: C

14. Colloids are best described as:

A. Large, heavy particles that settle quickly

B. Very small solids that stay dispersed due to size and electrical charge

C. Dissolved gases only

D. Any organism in water

Answer: B

15. Complete treatment, or conventional treatment, includes which steps?

A. Coagulant, sedimentation, and disinfection only

B. Coagulant, flash mix, coagulation-flocculation, sedimentation, and filtration

C. Screening, aeration, and disinfection only

D. Filtration and disinfection only

Answer: B

16. Direct filtration differs from complete treatment because it:

A. Does not use coagulant

B. Uses no filtration at all

C. Skips sedimentation and uses minimal flocculation

D. Only uses sedimentation

Answer: C

17. Direct filtration typically includes:

A. Coagulant chemicals, flash mixing, coagulation, minimal flocculation, and filtration

B. Coagulant chemicals, sedimentation, and filtration only

C. Aeration and disinfection only

D. Reverse osmosis only

Answer: A

18. The epilimnion is:

A. The bottom layer of a lake

B. The middle layer where temperature drops quickly

C. The upper, warmest layer of a stratified reservoir

D. A groundwater zone

Answer: C

19. The metalimnion is:

A. The surface mixing zone

B. The middle layer where temperature drops rapidly with depth

C. The deepest, coldest layer

D. The unsaturated zone above the water table

Answer: B

20. The hypolimnion is:

A. The warm, upper layer

B. The middle layer only in winter

C. The lowest, coldest layer in a stratified lake

D. The layer with the highest pH

Answer: C

21. Organic substances are those that:

A. Come only from minerals and rocks

B. Always contain nitrogen

C. Come from plant or animal sources and contain carbon

D. Do not contain carbon

Answer: C

22. Oxidation can be defined as:

A. Addition of hydrogen to a compound

B. Removal of oxygen from a compound

C. Addition of oxygen, removal of hydrogen, or loss of electrons

D. Lowering the pH

Answer: C

23. In water treatment, oxidation is commonly used to:

A. Increase turbidity

B. Convert iron and manganese into forms that can be filtered

C. Lower pH only

D. Make organics more soluble

Answer: B

24. pH is a measure of:

A. The hardness of water

B. The intensity of acidity or basicity of a liquid

C. Turbidity in water

D. Dissolved oxygen levels

Answer: B

25. A low pH, for example 5.5, indicates that the water is:

A. Highly basic, or alkaline

B. Neutral

C. Acidic

D. Free of organics

Answer: C

26. THM precursors are:

A. Inorganic hardness minerals

B. Natural organic compounds that can react with chlorine to form trihalomethanes

C. Only man-made chemicals

D. Bacteria that cause THMs

Answer: B

27. THM precursors are commonly found in:

A. Only groundwater

B. Only desalinated water

C. All surface waters and groundwaters as natural organics

D. Only treated water

Answer: C

28. One common result of algal bloom is:

A. Increased clarity and lower organics

B. Taste and odor problems in finished water

C. Lower pH and fewer organics

D. No effect on treatment plant operation

Answer: B

29. Which of the following is not a typical water quality problem caused by algal blooms?

A. Shorter filter runs

B. Increased pH

C. Dissolved oxygen depletion

D. Permanent increase in hardness

Answer: D

30. Algal blooms can lead to shorter filter runs because:

A. They reduce turbidity

B. Algae and organic matter plug filter media

C. Filters operate at lower flow rates

D. They remove silt from water

Answer: B

31. During an algal bloom, increased photosynthesis can cause:

A. Decreased pH

B. Increased pH

C. No change in pH

D. Decreased dissolved oxygen during the day

Answer: B

32. When large amounts of algae die and decompose, which problem is most likely in deeper water?

A. Increased dissolved oxygen

B. Dissolved oxygen depletion and possible anaerobic conditions

C. Increased hardness

D. Increased alkalinity

Answer: B

33. Which is not listed as a problem caused by algal blooms?

A. Taste and odor

B. Increased organic loading

C. Shorter filter runs

D. Decreased THM precursors

Answer: D

34. In thermal stratification, the warmest water is usually found in the:

A. Epilimnion

B. Metalimnion

C. Hypolimnion

D. Aquifer

Answer: A

35. In thermal stratification, the coldest water is usually found in the:

A. Epilimnion

B. Metalimnion

C. Hypolimnion

D. Surface microlayer

Answer: C

36. Anaerobic conditions in the hypolimnion can lead to sulfate reduction, producing:

A. Chlorine gas

B. Hydrogen sulfide, or H_2S

C. Carbon dioxide only

D. Ozone

Answer: B

37. Hydrogen sulfide in water is typically noticed by:

A. Metallic taste

B. Rotten egg odor

C. Blue color

D. No observable effect

Answer: B

38. Under anaerobic conditions near the reservoir bottom:

A. Iron and Manganese stay in solid form and settle

B. Iron and manganese go into solution

C. Iron and manganese are removed by filtration

D. Iron and manganese are converted to chlorine

Answer: B

39. When water with dissolved iron is chlorinated or aerated, a likely result is:

A. Brown or red stains on sinks and laundry

B. Strong chlorine taste only

C. Lower turbidity

D. Permanent softening

Answer: A

40. When water with dissolved manganese is oxidized, the typical customer complaint is:

A. White scale on fixtures

B. Black stains on sinks and clothing

C. Rotten egg odor

D. Bitter taste only

Answer: B

41. For algae control with copper sulfate, copper is most effective when alkalinity is:

A. 0–50 mg/L

B. 50–150 mg/L

C. Greater than 150 mg/L

D. Exactly 200 mg/L

Answer: A

42. For algae control with copper sulfate, when alkalinity is 50–150 mg/L, the operator should:

A. Not using copper sulfate at all

B. Use a lower dose

C. Increase the copper sulfate dose

D. Drain the reservoir

Answer: C

43. For algae control with copper sulfate, alkalinity greater than 150 mg/L means:

A. Treatment is most effective

B. Normal dose can be used

C. Copper sulfate application is not recommended

D. Only groundwater can be treated

Answer: C

44. Regarding pH and copper sulfate effectiveness for algae control:

A. Higher pH is better

B. Lower pH is better

C. pH has no effect

D. pH must be exactly 7.0

Answer: B

45. Copper sulfate for algae control works best when:

A. Suspended matter is high

B. Suspended matter is low

C. Turbidity is extremely high

D. The water is very muddy

Answer: B

46. High suspended solids during copper sulfate treatment tend to:

A. Increase copper availability

B. Have no effect

C. Bind copper and reduce its effectiveness

D. Remove all algae immediately

Answer: C

47. Organic matter and copper sulfate:

A. High organic matter makes copper more effective

B. Organic matter binds copper and lowers its effectiveness

C. Organic matter is not present in reservoirs

D. Organic matter only affects chlorine, not copper

Answer: B

48. For copper sulfate treatment, algae generally do not respond when water temperature is:

A. A. Above 80°F

B. Around 70°F

C. Below 50°F

D. Between 60–75°F

Answer: C

49. Temperature and copper sulfate effectiveness:

A. Higher temperature generally improves algae response to copper

B. Temperature has no effect

C. Lower temperature always improves effectiveness

D. Copper sulfate only works at freezing temperatures

Answer: A

50. When dosing copper sulfate for algae control, one important consideration is:

A. Only the flow rate

B. Only the chlorine residual

C. That different algae species respond differently to copper

D. That all algae respond the same to any dose

Answer: C

COAGULATION, FLOCCULATION, AND SEDIMENTATION

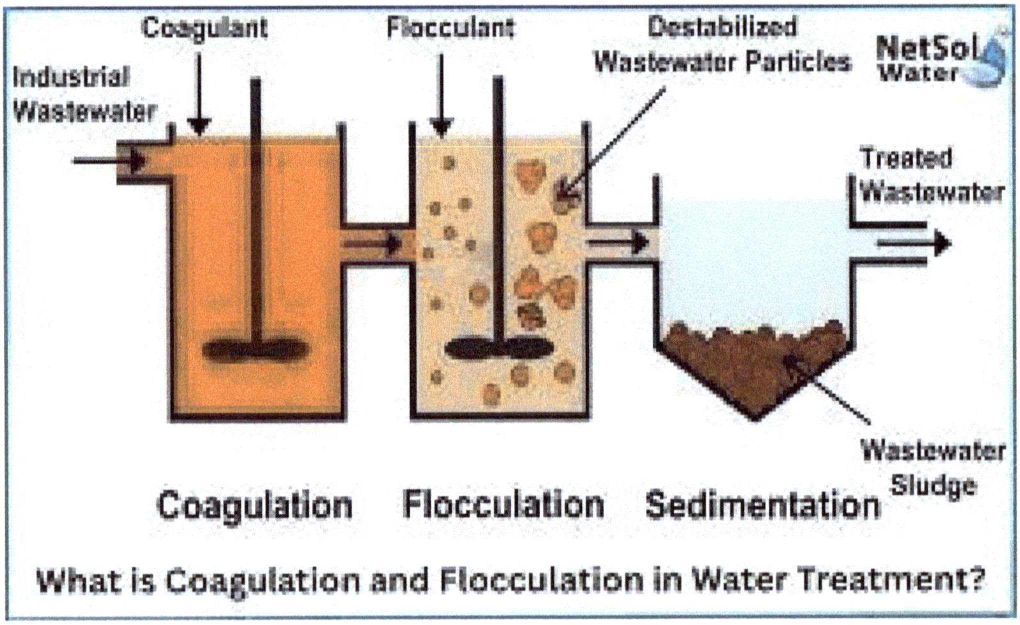

Alkalinity

Alkalinity is water's ability to resist pH change when acid is added. It mostly comes from carbonate, bicarbonate, and hydroxide, with smaller amounts from borate, silicate, and phosphate.

Apparent Color

Apparent color is the color you see in raw water with everything still in it, including dissolved substances and suspended particles, or turbidity.

True Color

True color is the color of water after turbidity is removed, usually by filtering, so only the dissolved substances causing color remain.

Coagulation

Coagulation is the process of adding chemicals, called coagulants, that neutralize particle charges so tiny particles start to stick together and form floc.

Floc

Floc is the soft, visible clumps made of particles and bacteria that have stuck together during coagulation and flocculation.

Flocculation

Flocculation is the slow, gentle mixing after coagulation that lets tiny coagulated particles collide and grow into larger, heavier floc that will settle or be filtered out.

Grab Sample

A grab sample is one single sample taken at one place and one time. It shows conditions only at that moment.

Composite Sample

A composite sample is made by combining many small samples taken at regular times over a set period, often 24 hours, so the final bottle represents the average conditions.

Representative Sample

A representative sample is a sample that matches the larger water body as closely as possible in quality and makeup, so test results truly reflect the system.

Disinfection Byproducts (DBPs)

DBPs are contaminants formed when a disinfectant, such as chlorine, reacts with other substances in the water, such as natural organic matter.

Jar Test

A jar test is a lab test that copies the plant's coagulation and flocculation process using several beakers with different chemical doses to find the best coagulant dose.

Specific Gravity

Specific gravity compares how heavy something is compared to the same volume of pure water.

- Water = 1.0 at 4°C.
- If SG > 1, it is heavier than water.
- If SG < 1, it is lighter than water.

CONVENTIONAL TREATMENT: 4 BASIC STEPS

- **Coagulation:** Add chemicals that neutralize particle charges so they can start to stick together.
- **Flocculation:** Gently mix the water so the small particles bump into each other and form larger floc.
- **Sedimentation:** Let the heavy floc settle to the bottom so it can be removed.
- **Filtration:** Pass the clarified water through filters to remove the remaining small particles.

COAGULATION CHEMISTRY BASICS

Most particles in water have a negative electrical charge.

A coagulant with positive ions is added to cancel, or neutralize, that negative charge.

Once the charges are neutralized, insoluble floc forms, which can then settle or be filtered.

The pH strongly affects how well the coagulant works. The wrong pH can cause poor coagulation.

If alkalinity is low, less than 40 mg/L, coagulation is harder, and alkalinity often must be added, such as lime or soda ash, to get good floc. Types of Chemicals Used as Coagulants

- Aluminum sulfate
- Ferrous sulfate
- Ferric sulfate
- Ferric chloride

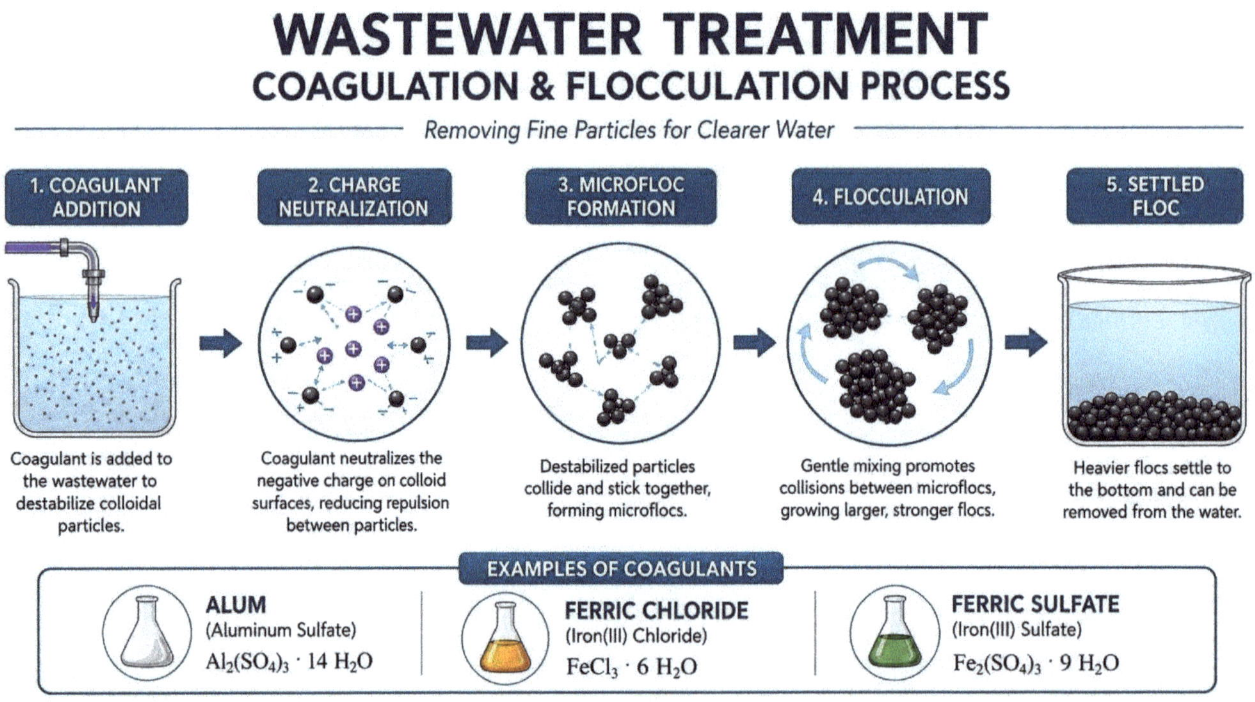

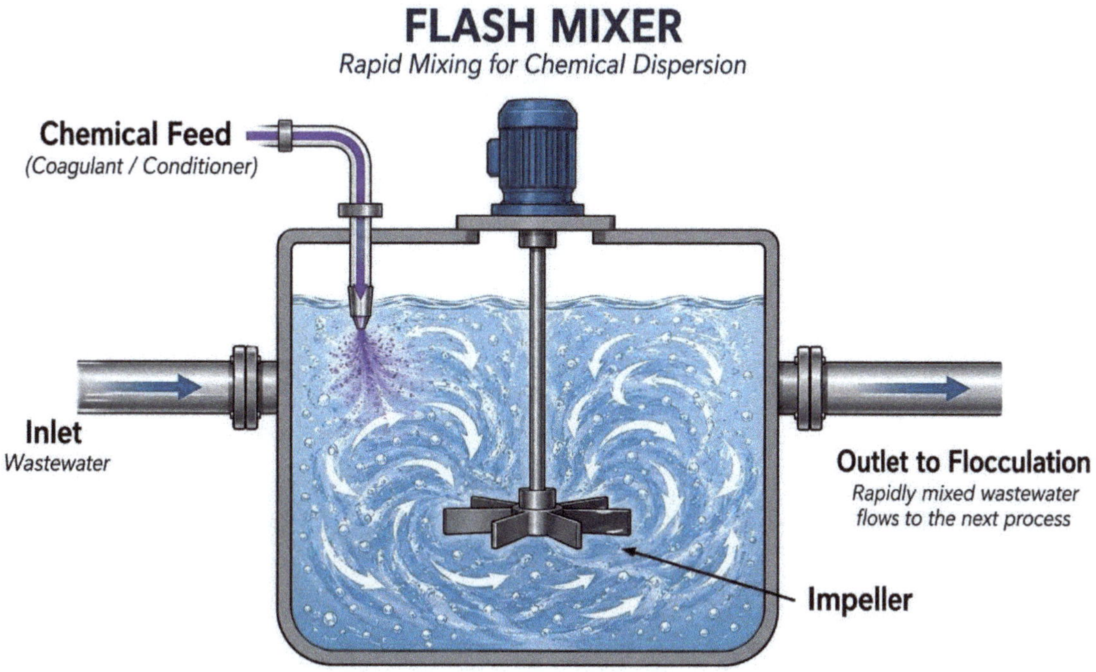

Types

A simple way to understand the four types of flash mixing:

- **Hydraulic mixing:** Uses the force of flowing water to mix chemicals.
- **Mechanical mixing:** Uses a motor and mixer blades to stir the water.
- **Diffuser:** Uses jets or nozzles to spread the chemical quickly into the water.
- **Pumped blender:** Uses a pump to move water and chemicals together so they mix quickly.

Why Optimum pH and Jar Testing Are Important in Coagulation and Flocculation

The goal is to help the chemical work better, use a lower dose, and make bigger, stronger floc.

Why Does pH Matter?

At the right pH, organic particles, such as humic and fulvic substances, become less dissolved and easier to remove. Lower pH can reduce how much coagulant is needed because the particles carry less charge and are easier to destabilize.

- Coagulant demand decreases.
- Flocculation improves at lower pH.
- Sulfuric acid can help before coagulant feed.

What Jar Testing Does?

Jar testing is a lab test used to find the best chemical dose and best pH for treatment. It helps operators avoid guessing and shows what works best for the actual water being treated.

Jar testing helps operators:

1. Find the best coagulant dose.
2. Find the best pH.
3. Find how much acid is needed.
4. See if alkalinity addition is needed.

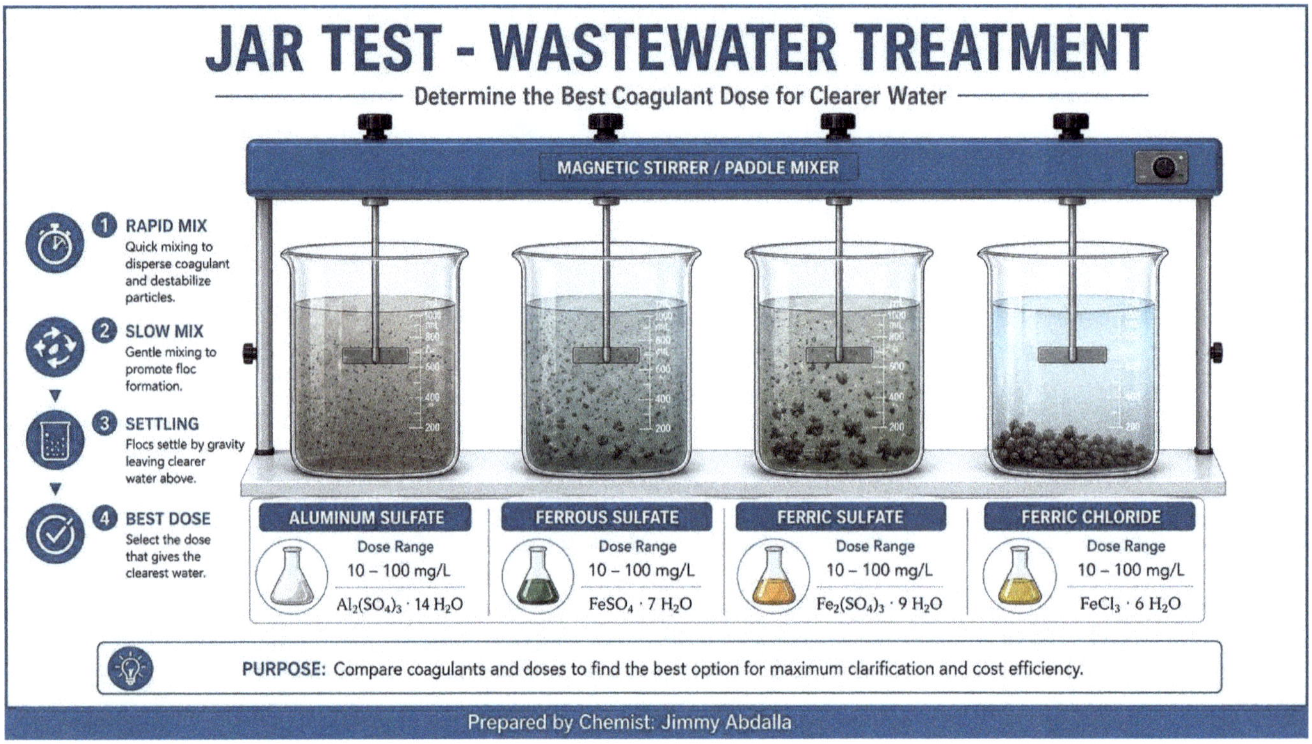

Because source water changes, the treatment process changes too.

When turbidity, pH, alkalinity, temperature, or color changes, the coagulant may not work the same way, so the floc can become weak or poor.

Why Source Water Changes Matter

- Turbidity changes how many particles need to be removed.
- pH affects how well the coagulant works.
- Alkalinity affects buffering and whether the water can hold the pH in the best range.
- Temperature affects reaction speed and floc formation.
- Color often means more dissolved organic matter, which can use up more coagulant.

Why Coagulation Changes

If raw water conditions shift, the same chemical dose may no longer work the same way. That can change pH, alkalinity, and turbidity removal, which then affects floc formation.

Why Floc Gets Poor

Poor floc forms when the particles are not properly destabilized or when the water chemistry is off. Common causes include wrong pH, low alkalinity, incorrect coagulant dose, or cold water slowing the process.

In simple words, the water changed, so the chemistry changed, and the floc got worse.

Coagulation and Flocculation in a Simple Way

Coagulation starts with rapid mixing to spread the chemical evenly. Then, flocculation uses gentle mixing to grow the floc. After that, the solids are removed by settling or filtration.

Coagulation means adding a chemical, such as alum or ferric chloride, to destabilize tiny particles in water. These particles are too small to settle on their own, so the chemical helps them lose their charge and start sticking together.

The first step is rapid mixing, also called flash mixing. This step spreads the coagulant through the water very quickly so every particle gets exposed to the chemical. Think of it like stirring sugar into tea quickly so it dissolves evenly.

After rapid mix, the water goes into flocculation, where the water is mixed slowly and gently. This lets the small clumps collide and grow into larger floc that can settle more easily. If the mixing is too strong, the floc can break apart.

Once the floc gets big enough, the water flows into a settling tank or clarifier. The heavy floc drops to the bottom and is removed as sludge. This step makes the water much clearer before filtration.

After settling, the water usually goes through filters, such as sand or membranes. Filtration removes any remaining fine particles that did not settle out.

Simple Memory Trick

You can remember the process as: add chemical, mix fast, mix slow, let settle, then filter.

Sedimentation

Sedimentation is when water is held still long enough for heavy particles to sink to the bottom. It helps remove solids before the water moves on to the next step.

Basic Treatment Steps

- Screening removes large debris, such as fish, sticks, and trash.
- Coagulation and flocculation add chemicals and mix the water so tiny particles stick together.
- Sedimentation lets those heavier particles settle to the bottom in a clarifier.
- Filtration removes particles that did not settle out.
- Disinfection kills bacteria and other disease-causing organisms.
- The clear well stores treated water and gives disinfectant contact time before delivery.

Sedimentation Terms

- **Clarifier:** A tank where solids settle to the bottom.
- **Influent:** Water coming into a tank or plant.
- **Effluent:** Water leaving a tank or plant.
- **Launders:** Channels that collect clarified water and carry it away.
- **Sedimentation:** The process of letting solids settle by gravity.

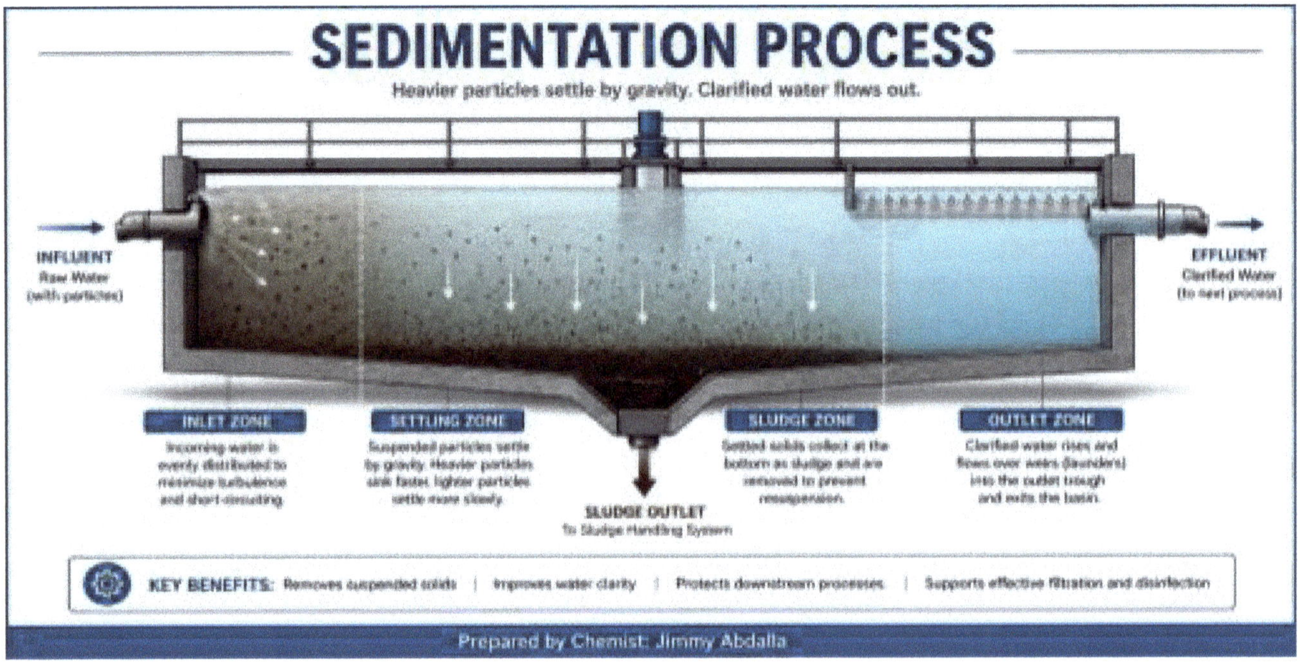

Important Ideas

- Density means how heavy something is for its size.
- Viscosity means how much water resists flowing. Cold water has higher viscosity.
- Plug flow means water moves through a tank in a fairly straight path.
- Short-circuiting means some water moves too quickly from the inlet to the outlet without enough treatment time.
- Overflow rate, or surface loading, shows how much water is being applied to the settling area.

Sedimentation Zones

A sedimentation basin usually has four zones: inlet zone, settling zone, sludge zone, and outlet zone. The goal is to keep the water calm long enough for solids to drop out.

Basin Types

Common sedimentation basins include rectangular, circular, square, double-decker, high-rate settlers, and solids-contact units. Different plants use different types depending on flow and design.

Performance

Good sedimentation can remove up to about 95% of suspended solids and often needs around 3 hours of detention time. If raw water is low in turbidity, some plants may use direct filtration instead.

Why Water Quality Changes Matter

If raw water turbidity, pH, alkalinity, temperature, or color changes, coagulation and flocculation can change too. That can cause poor floc, rising sludge, or worse settling in the basin.

A simple way to remember the full process is: screen, mix, floc, settle, filter, disinfect, store.

What are Launders?

Launders are channels that carry the cleaned water out of a sedimentation basin or filter. You can think of them as the “exit gutters” for treated water.

What is Overflow Rate?

Overflow rate tells how much water is being applied to the surface of the basin. It helps operators know if the clarifier is being overloaded.

What is Plug Flow?

Plug flow means water moves through the tank mostly in one direction, like a slug of water going from inlet to outlet. It is a good flow pattern because it gives water time to settle.

Short-Circuiting

Short-circuiting happens when some water takes a shortcut from the inlet to the outlet too fast. This reduces treatment time and makes the basin work poorly. In one sentence: sedimentation lets solids settle, launders carry water out, overflow rate shows loading, plug flow is the desired path, and short-circuiting is the problem to avoid.

What is Surface Loading?

Surface loading, also called overflow rate, tells how much water is being applied to the surface of a clarifier or settling tank. It helps operators see if the basin is carrying too much flow.

Simple idea: If the surface loading is too high, solids may not have enough time to settle. That can make the basin work poorly and let more solids escape.

What is Viscosity?

Viscosity is how much a liquid resist flowing. In simple words, it is how “thick” the water feels.

When water gets colder, its viscosity increases. That means cold water flows more slowly, and particles settle more slowly too.

Warm water is easier to move through a tank. Cold water is thicker and can make settling harder.

Easy Memory Trick

- Surface loading = how hard you are “loading” the basin with water.
- Viscosity = how thick or sticky the water is.

Sedimentation Performance

- Up to 95% REMOVAL of suspended solids means the basin can take out most of the dirt and particles from the water.
- About 3 hours of detention time means the water stays in the basin long enough for solids to settle by gravity.
- Max 2 NTU leaving the basin means the water should be fairly clear before it goes to the next step.
- Not necessary in low-turbidity waters means that if the source water is already very clean, the plant may skip sedimentation and use direct filtration instead.

Simple Meaning

Sedimentation works best when the water has enough time to rest and let particles settle. If the water is already low in turbidity, sedimentation may not add much value, so direct filtration can save time and cost.

In short, good sedimentation removes most solids, keeps water clear, and works best when the water quality is suitable for settling.

WHAT IS DIRECT FILTRATION?

Direct filtration is a water treatment process where water goes through coagulation, flocculation, and filtration without the sedimentation step. In simple words, the water is cleaned by chemicals and filters, but the settling tank is skipped.

When to Use It

Direct filtration is best for high-quality source water that is already fairly clean and stable. It is often used when raw water turbidity and color are low and when the water does not have heavy algae, iron, or manganese problems.

Simple Rule

A source of water may be a good candidate when turbidity is usually around 1 to 2 NTU, or generally low and consistent. If the water changes a lot or has many solids, conventional treatment with sedimentation is usually better.

Plants use direct filtration because it can reduce cost, space, sludge production, and chemical use compared with conventional treatment. The tradeoff is that it works best only when the raw water is stable and well behaved.

A simple way to remember it is direct filtration = no settling tank, just chemical mixing and filters.

CHAPTER 3 EXAM QUESTIONS

What is the main purpose of coagulation?

A. Kill bacteria

B. Remove hardness

C. Destabilize fine particles

D. Increase chlorine residual

Answer: C

What does flocculation do?

A. Settles sludge

B. Grows small particles into larger floc

C. Kills viruses

D. Removes taste and odor only

Answer: B

What is sedimentation?

A. Chemical addition

B. Particle settling by gravity

C. Filter backwashing

D. Chlorine contact

Answer: B

What is the main function of rapid mix?

A. Allow floc to settle

B. Distribute coagulant quickly

C. Filter suspended solids

D. Increase chlorine demand

Answer: B

What is the main function of slow mix?

A. Break floc apart

B. Grow floc by mixing gentle

C. Kill pathogens

D. Add alkalinity

Answer: B

What is a common primary coagulant?

A. Alum

B. Chlorine

C. Lime only

D. Carbon dioxide

Answer: A

What is used to determine the best coagulant dose?

A. pH meter

B. Jar test

C. Turbidimeter

D. Chlorine test

Answer: B

Coagulation works best when particles are mostly:

A. Positively charged

B. Neutral

C. Negatively charged colloids

D. Large sand grains

Answer: C

What is the waste from sedimentation called?

A. Effluent

B. Sludge

C. Residual ozone

D. Filter cake only

Answer: B

What is influent?

A. Water leaving a plant

B. Water entering a process

C. Sludge leaving a tank

D. Chemical residue

Answer: B

What is effluent?

A. Water entering a tank

B. Water leaving a tank or plant

C. Chemical added during rapid mix

D. Settled solids

Answer: B

WHAT IS floc?

A. Dissolved gas

B. Clumped particles

C. Chlorine residual

D. Filter media

Answer: B

What happens if flocculation mixing is too strong?

A. Floc grows faster

B. Floc breaks apart

C. Sedimentation improves

D. pH rises sharply

Answer: B

What does low water temperature usually do?

A. Speeds settling

B. Slows coagulation and flocculation

C. Removes color

D. Increases alkalinity

Answer: B

Why is alkalinity important in coagulation?

A. It disinfects water

B. It helps buffer pH changes

C. It increases turbidity

D. It removes iron directly

Answer: B

What is the common effect of too little alkalinity?

A. Better floc

B. pH drops too much

C. Higher chlorine residual

D. Faster settling

Answer: B

What is the purpose of coagulant aid?

A. Replace filtration

B. Improve floc formation

C. Increase hardness Kill algae

What is a common sign of poor coagulation?

A. Large, strong floc

B. Pin floc or cloudy water

C. Clear effluent

D. Stable settling

Answer: B

What is short-circuiting in a basin?

A. Water is moving too slowly

B. Water taking a direct path from inlet to outlet

C. Floc becoming too large

D. Sludge compaction

Answer: B

What does overflow rate measure?

A. Chlorine dose

B. Flow per surface area

C. Filter run time

D. Alkalinity demand

Answer: B

What does higher viscosity usually mean?

A. Water flows more easily

B. Water is “thicker” and moves more slowly

C. pH increases

D. More chlorine residual

Answer: B

Why is sedimentation important?

A. It adds chemicals

B. It removes settled solids before filtration

C. It disinfects water

D: changes color

Answer: D

one goal of sedimentation performance?

A. More turbidity after settling

B. Clearer water leaving the basin

C. Higher sludge in the outlet

D. Lower pH only

Answer: B

Which chemical is also called alum?

A. Aluminum sulfate

B. Sodium chloride

C. Calcium carbonate

D. Ferric nitrate

Answer: A

Ferric chloride is used mainly as a:

A. Disinfectant

B. Coagulant

C. Filter media

D. Corrosion inhibitor only

Answer: B

What is the main purpose of sedimentation basins?

A. Mix chemicals fast

B. Let solids settle

C. Increase chlorine demand

D. Raise pH

Answer: B

Which zone is found at the top of a sedimentation basin?

A. Sludge zone

B. Inlet zone

C. Outlet zone

D. Chemical zone

Answer: C

Which zone collects settled solids?

A. Inlet zone

B. Sludge zone

C. Outlet zone

D. D. Filter zone

Answer: B

What is one cause of poor settling?

A. Proper pH

B. Too much turbulence

C. Good detention time

D. Correct coagulant dose

Answer: B

What does a jar test help determine?

A. Filter media size

B. Best coagulant dose and pH

C. Chlorine contact time only

D. Pump horsepower

Answer: B

What happens when source water turbidity increases?

A. Less coagulants may be needed

B. More treatment may be needed

C. Sedimentation is unnecessary

D. Flocculation stops

Answer: B

What does color in source water often suggest?

A. Less organic matter

B. More natural organic material

C. No coagulation need

D. More chlorine only

Answer: B

Why is pH important in coagulation?

A. It controls filter size

B. It affects chemical performance

C. It replaces sedimentation

D. It removes all solids alone

Answer: B

is rising or floating sludge usually a sign of?

A. Excellent settling

B. Process problem

C. More disinfectants

D. Lower color

Answer: B

What is the main reason to use coagulants?

A. To dissolve solids

B. To help tiny particles stick together

C. Adding hardness

D. To make water cloudy

Answer: B

What is turbidity?

A. Water hardness

B. Cloudiness from suspended particles

C. Dissolved oxygen

D. Chlorine residual

Answer: B

What is a high-rate settler?

A. A disinfectant tank

B. A sedimentation unit that settles faster

C. A filter backwash tank

D. A Clearwell

Answer: B

What are a solids-contact unit?

A. A chlorination unit

B. A clarifier that helps floc contact solids

C. A filter underdrain

D. A sludge dryer

Answer: B

What is the main purpose of flocculation basins?

A. Kill bacteria

B. Increase particle collisions

C. Add chlorine

D. Remove hardness

Answer: B

What is one sign of hydraulic overload?

A. Lower flow than design

B. High surface loading

C. Better floc

D. Lower turbidity

Answer: B

What does detention time mean?

A. Time water stays in a tank

B. Time chlorine lasts in storage

C. Time a filter is cleaned

D. Time to collect samples only

Answer: A

Why is detention time important in sedimentation?

A. It gives solids time to settle

B. It increases algae growth

C. It raises pH

D. It destroys floc

Answer: A

What is one problem caused by short-circuiting?

A. Better settling

B. Less treatment time

C. More floc growth

D. Lower loading

Answer: B

What is a common test for checking settled water quality?

A. pH and turbidity

B. Chlorine smell only

C. Color chart only

D. Temperature only

Answer: A

What does alum consume during coagulation?

A. Chlorine

B. Alkalinity

C. Fluoride

D. Salt

Answer: B

Why might acid be added before coagulant?

A. To raise pH

B. To lower pH for better coagulation

C. To increase hardness

D. To stop sedimentation

Answer: B

What is surface loading also called?

A. Contact time

B. Overflow rate

C. Sludge age

D. Filter rate only

Answer: B

What does a higher surface loading rate mean?

A. Less water per area

B. More water per area

C. More alkalinity

D. Less turbidity

Answer: B

What is a clarifier another name for?

A. Sedimentation basin

B. Disinfection chamber

C. Chlorine contact tank

D. Filter gallery

Answer: A

What is the main goal of coagulation, flocculation, and sedimentation?

A. Make water cloudy

B. Remove suspended solids

C. Add dissolved solids

D. Increase odor

Answer: B

Why are colloids hard to remove?

A. They are too heavy

B. They carry charge and stay suspended

C. They sink too fast

D. They are filtered instantly

Answer: B

What helps neutralize colloid charge?

A. Coagulant

B. Chlorine

C. Fluoride

D. Soda ash only

Answer: A

What is a common sign that floc is forming well?

A. Tiny pin floc only

B. Large, visible floc

C. No particles

D. More color

Answer: B

Which is an example of physical treatment?

A. Flocculation

B. Disinfection

C. pH correction only

D. Taste and odor control

Answer: A

What is the usual effect of cold water on settling?

A. Faster settling

B. Slower settling

C. No change ever

D. Immediate filtration

Answer: B

the outlet zone in a sedimentation basin?

A. Where water enters

B. Where clarified water leaves

C. Where sludge enters

D. Where chemicals are stored

Answer: B

What happens if pH is not in the best range?

A. Coagulation may work poorly

B. Filtration becomes unnecessary

C. Chlorine disappears

D. Sludge vanishes

Answer: A

What is one reason plants use ferric salts?

A. They are filters

B. They act as coagulants

C. They are disinfectants

D. They soften water only

Answer: B

What is a typical result of good sedimentation?

A. Higher turbidity

B. Lower suspended solids

C. More algae

D. More short-circuiting

Answer: B

What happens if raw water quality changes a lot?

A. Same dose always works

B. Coagulation conditions may need adjustment

C. Sedimentation can be ignored

D. Flocculation is unnecessary

Answer: B

What is one reason to perform jar tests regularly?

A. To check only pipe size

B. To adjust to changing raw water

C. To reduce filter area

D. To increase sludge age

Answer: B

What is one role of sedimentation before filtration?

A. Remove the heaviest floc first

B. Add chlorine

C. Raise temperature

D. Create taste and odor

Answer: A

What is a low-turbidity water source more suited for?

A. Direct filtration

B. More sludge handling

C. More settling tanks

D. No treatment process

Answer: A

What does detention time depend on?

A. Tank volume and flow

B. Chlorine residual

C. Color only

D. Filter type only

Answer: A

What is the purpose of flocculation mixing speed control?

A. Keep floc from breaking

B. Increase short-circuiting

C. Stop coagulation

D. Remove chlorine

Answer: A

What is one thing operators should watch in a sedimentation basin?

A. Floc settling

B. Sludge buildup

C. Short-circuiting

D. All of the above

Answer: D

the best description of flocculation?

A. Rapid chemical mixing

B. Gentle mixing to build floc

C. Settling by gravity

D. Chlorine contact

Answer: B

What is the best description of coagulation?

A. Neutralizing particle charge

B. Drying sludge

C. Filtering water

D. Removing chlorine

Answer: A

What is a common coagulant aid?

A. Polymer

B. Sand

C. Chlorine gas

D. D. Air only

Answer: A

What does "surface loading" help determine?

A. Basin hydraulic overload

B. Filter media type

C. Chlorine demand

D. Sludge color

Answer: A

What is the main benefit of plug flow?

A. Better contact and settling conditions

B. More short-circuiting

C. Less treatment time

D. Higher turbidity

Answer: A

What is the problem if water goes through a basin too fast?

A. Overmixing

B. Short-circuiting

C. Better floc growth

D. Lower turbidity

Answer: B

What is the common effect of high-turbidity source water?

A. Less treatment needed

B. More coagulant demand

C. No floc needed

D. Lower loading automatically

Answer: B

What is the goal of the Clearwell?

A. Final settling of sludge

B. Disinfectant contact time and storage

C. Rapid mix

D. Coagulant mixing

Answer: B

What is one reason to lower pH in enhanced coagulation?

A. Help organics and particles destabilize

B. Increase color

C. Stop settling

D. Make water harder

Answer: A

Which is a common issue in a sedimentation tank?

A. Short-circuiting

B. Perfect plug flow always

C. No sludge production

D. No inflow

Answer: A

What does "max 2 NTU leaving basin" mean?

A. Water should be very clear after settling

B. Water should be dirty

C. Chlorine should be zero

D. pH should be 2

Answer: A

What should an operator adjust first if coagulation is poor?

A. Coagulant dose, pH, or mixing

B. Only the building lights

C. Only filter color

D. Only sludge pumps

Answer: A

92. What is the effect of increased water temperature on viscosity?

A. Viscosity increases

B. Viscosity decreases

C. No change

D. Water becomes solid

Answer: B

Why does low temperature make treatment harder?

A. Faster particle settling

B. Slower chemistry and settling

C. More chlorine production

D. Less turbidity always

Answer: B

What is one function of launders?

A. Store sludge

B. Collect and carry clarified water out

C. Add chemicals

D. Mix floc

Answer: B

What is the main purpose of a sedimentation basin?

A. Store chlorine

B. Let suspended solids settle out

C. Create turbulence

D. Increase color

Answer: B

What can cause settling problems even when the chemical dose is good?

A. Temperature changes

B. Correct pH

C. Proper mixing

D. Good detention time

Answer: A

What is one reason to monitor turbidity leaving the basin?

A. To verify clarification performance

B. To change color

C. To stop filtration

D. To increase sludge

Answer: A

What is the best summary of coagulation, flocculation, and sedimentation?

A. Add chemical, build floc, let solids settle

B. Chlorinate, filter, store

C. Heat, dry, pump

D. D. Mix, freeze, evaporate

Answer: A

ADSORPTION AND FILTRATION

Activated Carbon

Activated carbon is a black, porous carbon material used to remove certain chemicals, tastes, and odors from water. It works mainly by adsorption, which means contaminants stick to its surface. Simple Way to Think About It

Think of activated carbon like a tiny sponge with millions of pores. Water passes through it, and unwanted organic chemicals get trapped on the surface.

What It Removes

Activated carbon is especially good at removing:

- Chlorine
- Taste and odor compounds
- Many organic chemicals
- Some dissolved contaminants

Where It is Used

Water plants often use activated carbon after coagulation and sedimentation, or in filter systems, to improve water quality. It is also used in home water filters.

In short, activated carbon cleans water by grabbing unwanted chemicals onto its surface.

Adsorption is when dissolved substances in water stick to the surface of another material, such as activated carbon. It is a surface process, not something that soaks into the whole material.

Simple example: Adsorption is used to remove things like taste, odor, chlorine, organic chemicals, and some metals. Granular activated carbon is one common adsorptive medium.

Easy Memory Trick

- Adsorption = sticks on the surface.
- Absorption = goes into the material.

Backwash

Backwashing is cleaning a filter by reversing the water flow so trapped dirt and solids are washed out. It is done when the filter starts clogging or when the head loss gets too high.

Normal filtration pushes water down through the filter. Backwashing pushes clean water up through the filter so the dirt lifts out and goes to waste.

Backwashing helps:

- Remove trapped solids
- Lower pressure loss
- Restore filter performance
- Extend filter life

Easy memory trick: Filter gets dirty → reverse the flow → wash the dirt out.

Breakthrough

Breakthrough means particles start getting through a filter when they should be caught. In simple words, the filter is no longer holding everything back like it should.

Think of a filter like a net. At first, it catches the dirt, but if the filter gets too full, damaged, or overloaded, some particles start slipping through. That is breakthrough.

Breakthrough may happen when:

- The filter is saturated with solids.
- The media is damaged.
- The pressure is too high.
- The wrong filter size is used.

Breakthrough can cause dirty water to leave the filter and reduce water quality. Operators watch for it so they can backwash or fix the filter before performance gets worse.

Easy memory trick: Breakthrough = filter failure starts showing up.

Diatomaceous Earth Filter

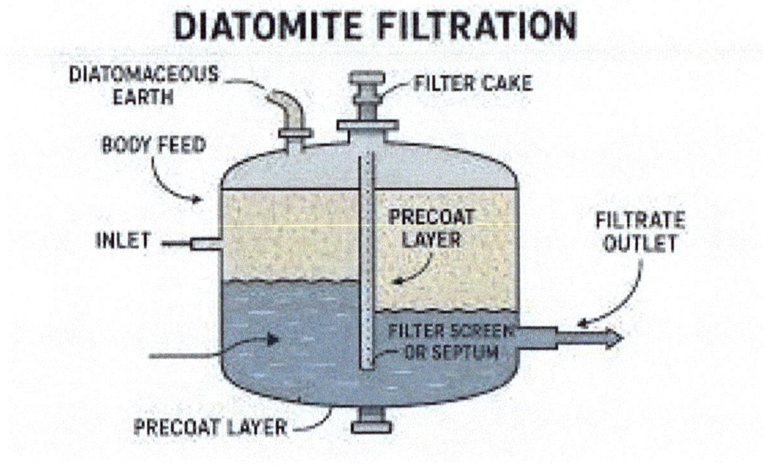

Diatomaceous earth filter, often called a DE filter, uses a fine powder made from the fossilized remains of tiny algae-like organisms called diatoms. In water treatment, it is used as filter media to trap very small particles.

Think of it like a very fine, porous screen. Water passes through the DE layer, and dirt and microorganisms get trapped.

DE filtration is effective for removing cysts, algae, and fine particles. It is often used in precoat filtration systems and is especially useful when very clear water is needed.

Easy Memory Trick

- Diatomaceous earth = fossil powder filter media.
- DE filter = fine filter that catches tiny particles.

Garnet

Garnet is a dense, hard mineral sand used as filter media in water treatment. It is usually the lowest layer in a multimedia filter because it is heavier than sand and stays in place during backwashing.

Garnet helps catch fine particles that pass through the upper filter layers. Think of it as the fine bottom layer that gives the filter extra polishing power.

Garnet is useful because it has high density, so it does not wash away easily. It is also hard and durable, so it lasts a long time. It improves filtration in multimedia filters and can help remove very small particles.

Easy memory trick: Garnet = heavy, fine, bottom filter media.

TYPES OF FILTRATION/FILTERS

1. Gravity Filtration

Gravity filtration means water passes through the filter by gravity only. It is common in rapid sand and multimedia filters, and the media can be single media, dual media, or multimedia.

Media types include:

- Single media: usually sand
- Dual media: sand and anthracite
- Multimedia: sand, anthracite, and garnet
- Activated carbon: used when adsorption is needed for taste, odor, or organics

Gravity filters usually run around 2 to 10 gpm/ft^2, with about 6 gpm/ft^2 as a normal maximum.

2. Pressure Filtration

Pressure filtration is similar to gravity filtration, but the filter is inside a closed tank. Water is pushed through under pressure, and the filtration rate is usually 2 to 3 gpm/ft^2.

3. Diatomaceous Earth Filtration

Diatomaceous earth, or DE, uses a fine powder made from diatoms as the filter medium. It is very good for catching very small particles and is often used when the water is already fairly clean.

4. Slow Sand Filtration

Slow sand filtration uses a very low filtration rate and a fine sand bed. The filtration rate is about 0.15 gpm/ft^2, and it is usually used for low-turbidity water.

Easy memory trick:

- Gravity = open filter, water flows by gravity.
- Pressure = closed tank, water is pushed through.
- DE = very fine powder filter.
- Slow sand = very slow, very gentle filtration.

HEAD LOSS

Head loss is the pressure drop that happens when water flows through a pipe or filter and meets resistance. In a filter, head loss gets bigger as dirt and solids build up in the media.

Think of it like trying to blow through a straw. A clean filter is easier to push water through, but a dirty filter takes more pressure because it is clogged.

Head loss can come from:

- Friction in pipes
- Rough pipe walls
- Fittings and bends
- Solids building up in a filter

As head loss increases, the filter may need backwashing. If it gets too high, water quality can get worse and breakthrough may happen.

Easy memory trick: Head loss = lost pressure. More dirt = more head loss.

WHAT IS THE UNIFORMITY COEFFICIENT?

Uniformity coefficient, or U.C., is a measure of how uniform the filter media particle sizes are. It compares the particle size at which 60% of the media is smaller with the particle size at which 10% of the media is smaller.

A lower U.C. means the media particles are more similar in size. A higher U.C.

means the media has a wider range of particle sizes.

The more uniform the media, the slower the head loss buildup. Media with a U.C. of 1.5 is commonly available. Media with a U.C. of 1.3 is also available, but it is usually more expensive.

Easy memory trick:

Lower U.C. = more uniform = better flow and slower clogging.

SPECIFIC GRAVITY

Specific gravity is a way to compare how heavy a substance is compared with the same volume of water.

Water is used as the reference, and its specific gravity is 1.000 at about 4°C. If something has a specific gravity greater than 1, it is heavier than water and tends to sink. If it is less than 1, it is lighter than water and tends to float.

Easy example: Rock has a specific gravity above 1, so it sinks. Oil has a specific gravity below 1, so it floats.

Exam-style definition:

Specific gravity = the weight or density of a substance compared to the weight of the same volume of water.

SOFTENER, SOFTENING, AND HARDNESS

Softener = hardness remover.

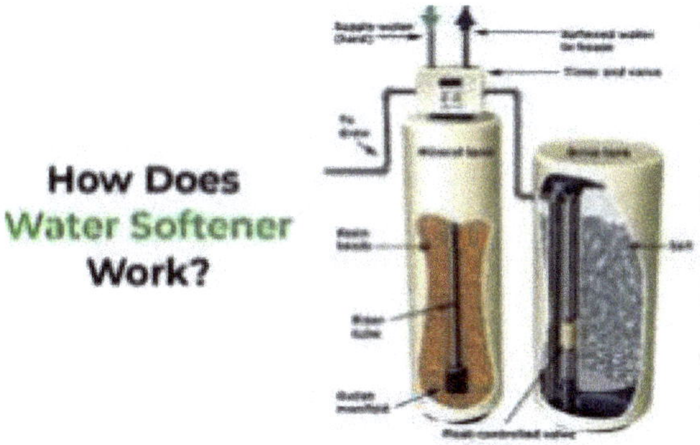

It is a water treatment unit that removes hardness from water. It usually does this by ion exchange, where calcium and magnesium are swapped for sodium or potassium.

Hard water goes into the softener, passes through resin beads, and comes out softer.

The resin beads hold the hardness minerals until they are regenerated with brine.

Soft water helps reduce scale in pipes and equipment and makes soap work better. That is why softeners are common in homes and treatment systems.

Hardness

Water Hardness Scale		
mg/L & ppm as CaCO3	dH	Classification
0 - 75	0 - 4	Soft
75 -150	4 - 8.5	Slightly Hard
150 - 300	8.5 - 17	Hard
300+	17+	Very Hard

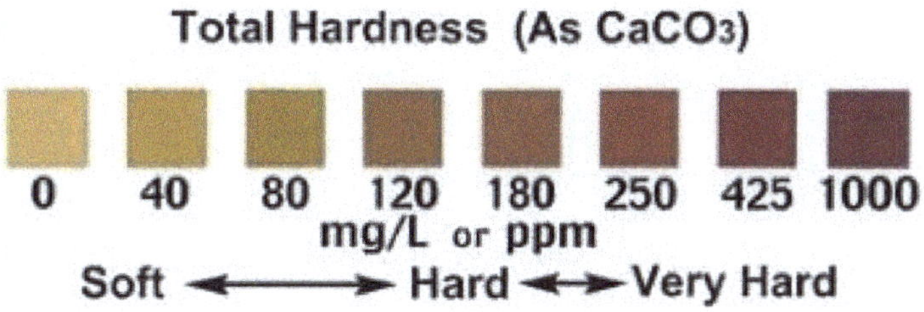

Hardness is the amount of dissolved calcium and magnesium in water. Hard water can make soap less effective and can cause scale in pipes and equipment.

If water has a lot of calcium and magnesium, it is called hard. If it has very little, it is soft.

Hardness can cause:

- Scale buildup
- Soap not lathering well
- Operational problems in equipment

Exam-style definition: Hardness = the amount of dissolved calcium and magnesium in water.

GIARDIA AND VIRUSES

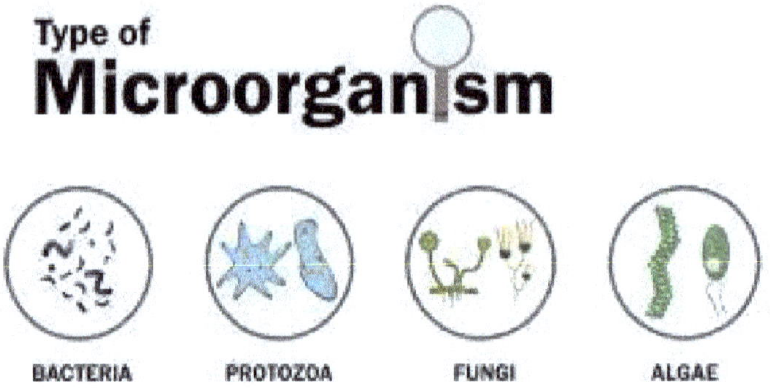

Giardia and viruses are two important microorganisms that water treatment must control. For surface water systems, the usual goal is 99.9%, or 3-log, removal or inactivation of Giardia and 99.99%, or 4-log, removal or inactivation of viruses. A 3-log reduction means the organism is reduced by 99.9%. A 4-log reduction means it is reduced by 99.99%. In simple terms, the higher the log number, the greater the removal or kill rate.

Giardia is a parasite that can cause stomach illness if it gets into drinking water. Treatment plants remove it through filtration and inactivate it through disinfection, and the combined process must meet the 3-log target.

Viruses are even smaller than Giardia and are harder to remove physically, so strong disinfection is very important. The required target is 4-log removal or inactivation, which means 99.99%.

How Plants Achieve It

- Filtration helps remove Giardia and some viruses.
- Disinfection provides additional inactivation, often measured by CT or UV dose. • Good plant operation is important because poor filtration or low disinfectant contact time can reduce treatment credit.

Easy memory trick:

- Giardia = 3-log = 99.9%
- Virus = 4-log = 99.99%

TURBIDIMETER AND PARTICLE COUNTER

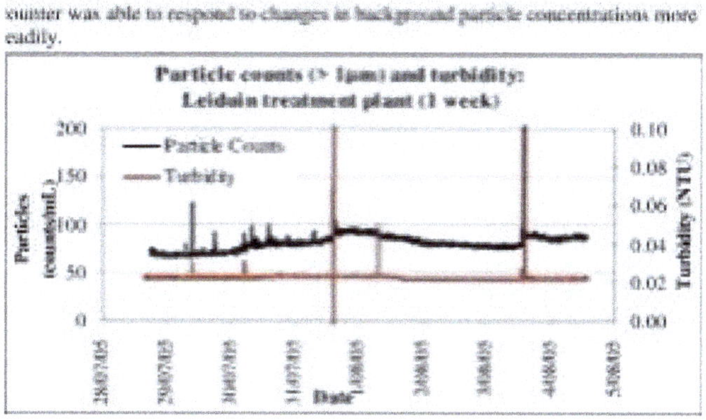

Turbidimeter

A turbidimeter measures how much light is scattered by particles in water, which tells you how cloudy the water is. It does not tell you the number of particles or the exact particle size.

One large particle can give about the same turbidity as many small particles, so turbidity alone can hide what is really happening in the water.

Particle Counter

A particle counter measures the number of particles and can also sort them by size. That makes it more useful when you want to know what sizes are getting through a filter.

Particle counters are especially helpful for monitoring filter effluent because they can detect particles in the size range that matters for Cryptosporidium and similar organisms. They give a more sensitive picture than turbidity when finished water is very clean.

Simple memory trick:

- Turbidity = how cloudy the water looks.
- Particle counter = how many particles are there and how big they are. In one sentence: turbidity tells you cloudiness, while particle counting tells you particle number and size, so particle counters are better for catching filter problems early.

CHAPTER 4 EXAM QUESTIONS

1. What is the main purpose of activated carbon in water treatment?

A. Add hardness

B. Remove taste and odor

C. Increase turbidity

D. Raise pH

Answer: B

2. Activated carbon works mainly by:

A. Sedimentation

B. Adsorption

C. Coagulation

D. Chlorination

Answer: B

3. Head loss in a filter means:

A. Increase in water color

B. Pressure drop through the filter

C. Increase in chlorine demand

D. Decrease in turbidity only

Answer: B

4. Backwashing is used to:

A. Add chemicals to the water

B. Reverse flow and clean the filter

C. Increase hardness

D. Remove chlorine

Answer: B

5. Breakthrough means:

A. A filter is brand new

B. Particles begin passing through the filter

C. pH becomes neutral

D. A pump stops running

Answer: B

6. Water softening removes:

A. Nitrate

B. Calcium and magnesium

C. Chlorine

D. Fluoride

Answer: B

7. A water softener works by:

A. Sedimentation

B. Ion exchange

C. Oxidation

D. Coagulation

Answer: B

8. In ion exchange softening, hardness ions are replaced mainly by:

A. Chloride

B. Sodium

C. Iron

D. Sulfate

Answer: B

9. Which mineral is commonly used as filter media in some filtration systems?

A. Activated carbon

B. Salt

C. Lime

D. Soda ash

Answer: A

10. The purpose of backwashing is to:

A. Compact the bed

B. Remove trapped solids

C. Increase hardness

D. Reduce flow to zero forever

Answer: B

11. A clean filter usually has:

A. High head loss

B. Low head loss

C. No flow

D. High hardness

Answer: B

12. As a filter gets dirty, head loss usually:

A. Decreases

B. Increases

C. Stays the same

D. Becomes zero

Answer: B

13. Adsorption means contaminants:

A. Dissolve into the whole media

B. Stick to the surface of a material

C. Turn into gas

D. Sink to the bottom

Answer: B

14. The term "hard water" refers mainly to water with high:

A. Sodium and chloride

B. Calcium and magnesium

C. Iron and manganese only

D. Fluoride and nitrate

Answer: B

15. Which is a sign that a filter may need backwashing?

A. Lower head loss

B. Higher head loss

C. Less media

D. More sodium

Answer: B

16. A filter breakthrough can affect:

A. Water quality leaving the filter

B. Only pump speed

C. Salt usage only

D. Tank color only

Answer: A

17. In a softener, resin beads:

A. Add hardness ions

B. Swap ions with hardness minerals

C. Remove chlorine by oxidation

D. Kill bacteria by UV light

Answer: B

18. A filter that is too full of solids may:

A. Have lower pressure drop

B. Have higher pressure drop

C. Stop adsorbing immediately

D. Become a softener

Answer: B

19. One purpose of filtration is to:

A. Increase taste and odor

B. Remove suspended particles

C. Add hardness

D. Raise sodium

Answer: B

20. Which process is most closely associated with a deep bed filter?

A. Backwashing

B. Fermentation

C. Freezing

D. Chlorination

Answer: A

21. What is the main job of filter media?

A. Store chlorine

B. Trap solids and particles

C. Add alkalinity

D. Lower temperature

Answer: B

22. Activated carbon is a type of:

A. Chemical coagulant

B. Adsorptive media

C. Disinfectant

D. Lime softener

Answer: B

23. A water softener does not mainly remove:

A. Hardness

B. Calcium

C. Magnesium

D. Sand

Answer: D

24. A filter with low head loss and good performance is usually:

A. Clean

B. Clogged

C. Broken

D. Overloaded

Answer: A

25. What happens during regeneration of a softener?

A. Resin is heated

B. Brine restores the resin

C. Chlorine is added

D. Media is dried

Answer: B

26. Which is an example of a treatment barrier?

A. Softening

B. Logging

C. Flushing only

D. Screening only

Answer: A

27. Log removal is used to describe:

A. Pressure loss in a filter

B. Reduction of microorganisms by treatment

C. Salt dosage

D. Turbidity meter calibration

Answer: B

28. If particles begin passing through a filter, the operator should think about:

A. Breakthrough

B. Free chlorine residual

C. Pump cavitation

D. Softening

Answer: A

29. Hardness in water is often measured by:

A. Calcium carbonate equivalent

B. Sodium chloride equivalent

C. Chlorine demand

D. BOD

Answer: A

30. The filter cleaning process that reverses flow is:

A. Coagulation

B. Backwashing

C. Aeration

D. Flocculation

Answer: B

31. Soft water contains less:

A. Calcium and magnesium

B. Chlorine and fluoride

C. Oxygen and nitrogen

D. Sand and silt

Answer: A

32. Filter clogging usually causes:

A. Less head loss

B. More head loss

C. More sodium removal

D. Better adsorption forever

Answer: B

33. Activated carbon is often used after:

A. Filtration only

B. Coagulation and sedimentation

C. Lime softening only

D. Backwashing only

Answer: B

34. A softener is most useful when water has:

A. High hardness

B. High chlorine only

C. High color only

D. High fluoride only

Answer: A

35. The best reason to backwash a filter is to:

A. Raise pressure loss

B. Remove trapped solids and restore flow

C. Add hardness ions

D. Increase turbidity

Answer: B

36. Which process uses resin beads?

A. Sedimentation

B. Ion exchange

C. Perchlorinating

D. Aeration

Answer: B

37. Which ion is commonly exchanged for hardness ions in a softener?

A. Sodium

B. Iron

C. Aluminum

D. Silica

Answer: A

38. A filter media bed becomes dirty when it:

A. Traps solids

B. Adds sodium

C. Raises chlorine residual

D. Makes water harder

Answer: A

39. Head loss is best described as:

A. Water color

B. Resistance to flow

C. Hardness removal

D. Disinfection power

Answer: B

40. The main role of activated carbon is to:

A. Remove dissolved taste and odor compounds

B. Add alkalinity

C. Remove sand only

D. Kill all bacteria

Answer: A

41. A clean bed in filtration means:

A. A bed with no media

B. Media with low trapped solids

C. A softener regenerated

D. A tank with no flow

Answer: B

42. Which is most likely to be removed by activated carbon?

A. Chlorine

B. Calcium hardness

C. Sodium hardness

D. Gravel

Answer: A

43. A softener is regenerated with:

A. Brine

B. Carbon dioxide

C. Alum

D. Fluoride

Answer: A

44. The type of process that removes hardness by replacing ions is:

A. Ion exchange

B. Decantation

C. Screening

D. Adsorption

Answer: A

45. Breakthrough is a problem because:

A. Water becomes softer

B. Contaminants start getting through

C. Head loss goes to zero

D. Media becomes magnetic

Answer: B

46. What does a turbidimeter measured?

A. Particle size and count

B. Light scattering caused by cloudiness

C. Hardness ions

D. Salt concentration

Answer: B

47. Turbidity alone does not tell you:

A. Whether water is cloudy

B. Number or size of particles

C. Light scattering

D. Filter color

Answer: B

48. Particle counters are more useful than turbidity for:

A. Measuring hardness

B. Monitoring fine particles in filter effluent

C. Adding chlorine

D. Increasing pressure

Answer: B

49. In a softener, the resin bed eventually becomes:

A. Saturated with hardness ions

B. Turned into sand

C. Disinfected

D. Evaporated

Answer: A

50. Which is a reason to monitor filter effluent?

A. To check for particle breakthrough

B. To increase salt use

C. To lower temperature

D. To reduce media size

Answer: A

51. A filter media bed that is overloaded may:

A. Need backwashing sooner

B. Need more hardness

C. Stop flowing forever

D. Become a softener

Answer: A

52. The term "hardness removal" is most closely tied to:

A. Ion exchange softening

B. Adsorption

C. Sedimentation

D. Aeration

Answer: A

53. Which statement best fits adsorption?

A. Contaminants stick to surface pores

B. Minerals settle by gravity

C. Water is boiled

D. Resin adds hardness

Answer: A

54. Which process is used to clean most granular filters?

A. Backwashing

B. Chlorination

C. Softening

D. Coagulation

Answer: A

55. The main difference between turbidity and particle counting is:

A. Turbidity measures light scattering, while particle counters measure number and size

B. They are exactly the same

C. Particle counters only measure color

D. Turbidity measures hardness

Answer: A

56. A dirty filter usually produces:

A. Lower head loss

B. Higher head loss

C. More softening

D. Less flow resistance

Answer: B

57. What does regeneration do in a softener?

A. Recharges the resin

B. Removes chlorine

C. Adds sand

D. Measures turbidity

Answer: A

58. Which is a likely result of breakthrough?

A. Better water quality

B. More particles in the effluent

C. Lower hardness

D. Lower flow in a pipe only

Answer: B

59. Hard water is mainly caused by:

A. Calcium and magnesium

B. Sodium and chloride

C. Iron and aluminum

D. Fluoride and nitrate

Answer: A

60. Which statement about filter media is correct?

A. It traps solids as water passes through

B. It adds hardness

C. It is only used in softeners

D. It makes water cloudy

Answer: A

61. Activated carbon is best known for removing:

A. Taste and odor compounds

B. Calcium carbonate only

C. Sand and gravel only

D. Hardness ions only

Answer: A

62. What is the purpose of a filter effluent check?

A. Find if particles are passing through

B. Measure tank color

C. Add alum

D. Raise pH only

Answer: A

63. Which operation restores filter capacity after solids build up?

A. Backwash

B. Sedimentation

C. Softening

D. Coagulation

Answer: A

64. Log removal is important because it shows:

A. How many microorganisms are reduced

B. How much sodium is added

C. How hard the water is

D. How fast the pump turns

Answer: A

65. Which is not a softening job?

A. Removing Ca and Mg

B. Preventing scale

C. Increasing hardness

D. Ion exchange

Answer: C

66. Head loss rises as:

A. Solids accumulate in the filter

B. Water gets softer

C. Chlorine rises

D. Resin recharges

Answer: A

67. A good filter should:

A. Remove particles efficiently

B. Add particles

C. Increase turbidity

D. Stop flow completely

Answer: A

68. What is one reason to use activated carbon after treatment?

A. Improve taste and odor

B. Increase hardness

C. Make more solids

D. Replace filtration

Answer: A

69. Which is a common sign that a softener needs regeneration?

A. Resin is exhausted with hardness

B. Head loss in a filter is low

C. pH is always 7

D. Water is too cold

Answer: A

70. A filter that is "clean" usually has:

A. Lower pressure drop

B. Higher pressure drop

C. More breakthrough

D. More hardness

Answer: A

71. Which process is surface-based?

A. Adsorption

B. Softening

C. Sedimentation

D. Flocculation

Answer: A

72. What does "log removal" describe?

A. Pressure loss in a filter

B. Reduction by powers of ten

C. Salt dose in a softener

D. Turbidity units

Answer: B

73. Particle counters are useful because they:

A. Measure particle number and size

B. Measure hardness only

C. Remove chlorine

D. Measure pH only

Answer: A

74. Turbidity can miss:

A. Small changes in particle count

B. Filter backwash

C. Hardness ions

D. Resin beads

Answer: A

75. Which media is porous and adsorptive?

A. Activated carbon

B. Gravel

C. Sand only

D. Lime

Answer: A

76. In softening, sodium is usually:

A. Released into the water

B. Removed by sedimentation

C. Turned into hardness

D. Filtered by sand

Answer: A

77. The purpose of filter backwashing is to:

A. Clean trapped solids from the media

B. Add more solids

C. Increase hardness

D. Stop flow permanently

Answer: A

78. Which of these is most related to filter clogging?

A. Head loss

B. pH

C. Sodium dose

D. Water color only

Answer: A

79. A water softener helps reduce:

A. Scale

B. Particle count only

C. Chlorine residual

D. Filter turbidity only

Answer: A

80. Breakthrough is most concerning because it means:

A. Treatment is failing to hold contaminants

B. Hardness is lower

C. Backwash is complete

D. Resin is new

Answer: A

81. What is the main function of resin beads?

A. Swap ions during softening

B. Hold sand in place

C. Add chlorine

D. Increase turbidity

Answer: A

82. Which process is used to remove taste and odor compounds?

A. Activated carbon adsorption

B. Sedimentation

C. Lime softening

D. Screening

Answer: A

83. Which of the following is a filtration problem?

A. High head loss

B. Lower hardness

C. Better taste

D. Soft water

Answer: A

84. Which is a softening media?

A. Ion exchange resin

B. Anthracite only

C. Sand only

D. Gravel only

Answer: A

85. A particle counter is more useful than turbidity when monitoring:

A. Fine filter effluent particles

B. Salt brine strength

C. Pipe corrosion only

D. Water color only

Answer: A

86. What does “adsorption” mean in activated carbon treatment?

A. Contaminants stick to the surface

B. Contaminants settle

C. Contaminants freeze

D. Contaminants become hardness

Answer: A

87. Soft water is less likely to:

A. Form scale

B. Flow through pipes

C. Enter a home

D. Have any ions

Answer: A

88. Which statement is true?

A. Turbidity measures particle size and count

B. Particle counters can size and count particles

C. Softeners remove chlorine

D. Activated carbon adds hardness

Answer: B

89. A softener regenerates with:

A. Salt brine

B. Gravel

C. Alum

D. Carbon dioxide

Answer: A

CHAPTER 5

DISINFECTION

What Disinfection Does

Disinfection kills or inactivates disease-causing organisms in water so the water is safe to drink. It is a treatment barrier, not just a cleaning step, because it reduces microbial risk before the water reaches customers.

Main Disinfectants

Common disinfectants include chlorine, chloramine, ozone, and ultraviolet light. For T2, chlorine is the most important because operators can measure a residual and use it for compliance and process control.

Chlorine Residual

Free chlorine is the usable disinfectant remaining in the water after demand is satisfied. If the residual is too low, the system may not have enough disinfection power.

CT and Contact Time

CT means concentration times contact time, or **C** × **T**. A higher residual or longer contact time increases disinfection effectiveness, and the two work together.

Baffling and Clearwells

Contact time is usually provided in a clearwell or contact tank. Baffling improves effective contact time by making water travel a longer and more uniform path, while poor baffling lowers treatment efficiency.

What Affects Chlorine

Chlorine works better at lower pH and higher temperature. Chlorine demand, organic matter, and short contact time can reduce how well disinfection works.

Log Removal

Log removal is a way to describe pathogen reduction in powers of ten. A 1-log reduction is 90%, 2-log is 99%, 3-log is 99.9%, and 4-log is 99.99%.

Treatment Train Idea

Filtration and disinfection work together as barriers in the treatment train. Filtration removes particles, while disinfection inactivates microorganisms that remain.

Operator Checks

Operators monitor chlorine residual, contact time, pH, temperature, and system performance to confirm disinfection is adequate. A low residual or poor CT can mean the system needs more chlorine, more contact time, or better baffling.

Simple Exam Facts

A common exam idea is that CT is the product of chlorine residual and contact time. Another common idea is that if the contact time is too short, the disinfection process may be incomplete even when chlorine is present.

Quick Example

If free chlorine residual is 1.0 mg/L and contact time is 10 minutes, then CT is 10 mg-min/L. That simple calculation is the kind of question often tested on operator exams.

UV

UV disinfection uses ultraviolet light, usually UV-C around 254 nm, to damage the DNA or RNA of microorganisms so they cannot reproduce or cause infection. It is a physical disinfection process, not a chemical one, so it does not add disinfectant residual to the water.

Water flows through a UV reactor or chamber where it passes around UV lamps. The UV light inactivates bacteria, viruses, and some protozoa by disrupting their internal genetic material.

Main Advantages

UV is fast, chemical-free, and does not change taste, odor, pH, or temperature. It is also effective against organisms that can be hard to control with chlorine, including Giardia and Cryptosporidium under proper design conditions.

Limitations

UV does not remove sediment, hardness, or chemical contaminants, and it provides no disinfectant residual in the distribution system. Water quality matters because particles, iron, manganese, or turbidity can block or scatter UV light and reduce effectiveness.

What Operators Should Remember

The effectiveness of UV depends on dose, intensity, and exposure time. If the water is too dirty or the UV system is undersized, disinfection may be incomplete.

Simple Memory Trick

UV means light in and microbes out of business. It damages DNA or RNA so they cannot reproduce.

OZONE, OR O_3

O_3 means ozone. In water treatment, ozone is a strong oxidant and disinfectant that is used to inactivate microorganisms and oxidize taste-and-odor compounds, iron, manganese, and some organics.

Ozone is generated on-site, usually with an ozone generator, and then injected into water through a contact chamber or diffuser system. Once dissolved, it reacts quickly with contaminants and microbes. The remaining ozone is usually destroyed or allowed to decompose back to oxygen.

WHY OPERATORS LIKE OZONE

Ozone acts fast, has strong oxidation power, and can reduce taste, odor, iron, manganese, and microbial contamination. It also leaves little to no lasting chemical residual in the water because it breaks down to oxygen.

Ozone does not provide the same lasting distribution-system residual that chlorine does, so it is often paired with another disinfectant, such as chlorine or chloramine.

CHLORINE VS. OZONE

Ozone is a very strong oxidant and disinfectant that works fast and is effective against many pathogens. It also helps oxidize taste, odor, iron, manganese, and some organic compounds. The big drawback is that it does not leave a lasting residual in the distribution system.

Chlorine is widely used because it is effective, inexpensive, and leaves a measurable residual in the water. That residual protects the water as it travels through pipes and storage tanks. A drawback is that chlorine can form disinfection byproducts, such as THMs and HAAs.

CHLORINE CURVE OR BREAKPOINT CHLORINATION

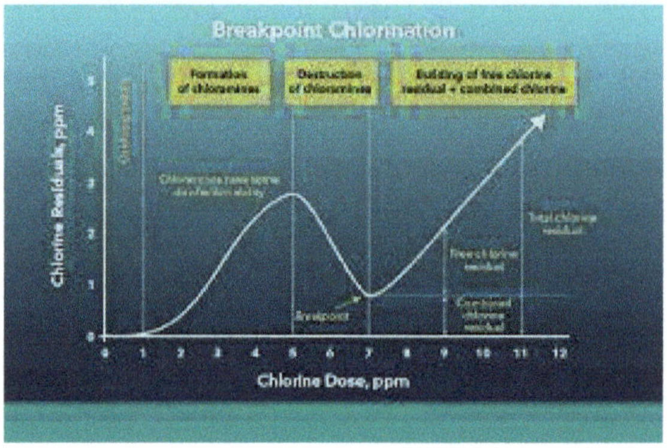

The chlorine curve

usually means the breakpoint chlorination curve. It shows how chlorine residual changes as more chlorine is added to water, especially when ammonia or other chlorine-demanding substances are present.

At first, added chlorine is used up by chlorine demand, so there may be little or no free residual.

Then chloramines form, and the residual may rise and fall as chlorine keeps reacting. At the breakpoint, the demand is satisfied, and the chlorine residual starts increasing again as free chlorine.

WHY OPERATORS CARE

The curve helps operators find the right chlorine dose to overcome demand and maintain a disinfectant residual. It is used to control disinfection, minimize chloramines when desired, and make sure the water stays protected. Simple Memory Trick

Chlorine goes in, demand gets satisfied, breakpoint is reached, and then free chlorine remains.

CHLORAMINES

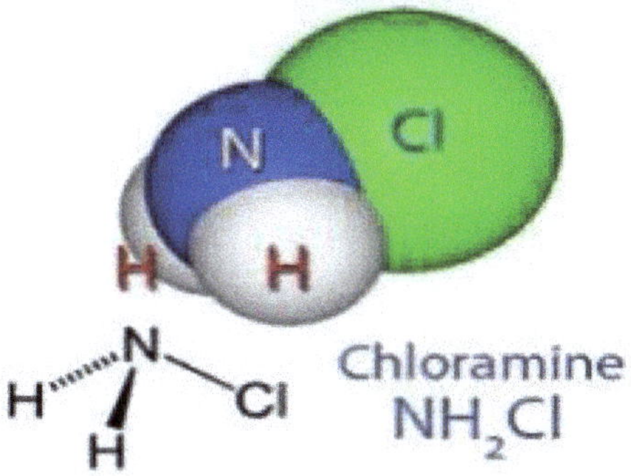

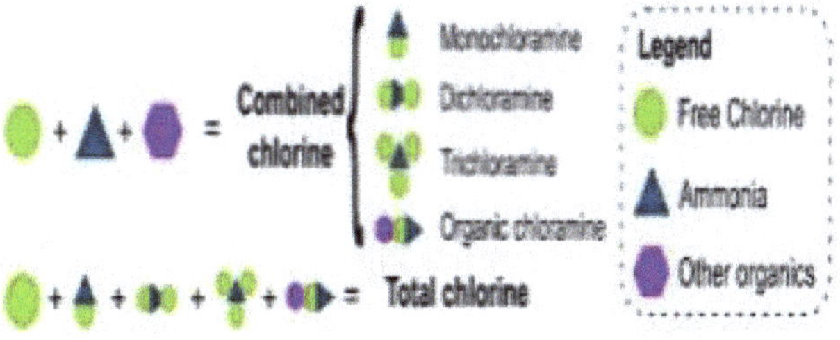

Chloramines are disinfectants formed when chlorine combines with ammonia. The main species desired for drinking water treatment is monochloramine because it is more stable and useful for maintaining a residual in the distribution system.

Many utilities use chloramines as a secondary disinfectant because they last longer in pipes than free chlorine. That longer residual helps protect water farther from the plant and can reduce the formation of some chlorine-related disinfection byproducts.

Chloramines are made by carefully adding ammonia to chlorinated water or by combining chlorine and ammonia in a controlled way so the chlorine-to-ammonia ratio favors monochloramine. If the ratio is not controlled well, other forms, such as dichloramine or trichloramine, can form. These are less desirable because they can create taste and odor problems.

TYPES OF CHLORAMINES

There are three main chloramine species: monochloramine, dichloramine, and trichloramine.

Monochloramine is the target disinfectant in drinking water, while the others are generally less desirable because they can worsen taste, odor, and water quality complaints.

Chloramines are weaker and slower acting than free chlorine, but they persist longer in the distribution system. That makes them useful where a long-lasting residual is more important than rapid disinfection.

A key advantage of chloramines is that they can provide a stable disinfectant residual in the distribution system. This is one reason systems may switch from free chlorine to chloramination, especially when they want better residual maintenance or fewer chlorine byproducts. The biggest operational problem with chloramines is nitrification. Nitrification is a biological process where bacteria use ammonia and chloramine decay products. This can reduce chloramine residual, create nitrite and nitrate, and cause water quality problems.

When nitrification happens, chloramine residual drops, free ammonia can change, and nitrite may rise. This can lead to odors, loss of disinfectant protection, and the need for operational changes, such as flushing, booster chlorination, or better ammonia control.

Operators try to keep free ammonia low and maintain the proper chlorine-to-ammonia balance. Some systems use booster chlorine to recombine leftover ammonia into chloramines before nitrification becomes a problem.

<u>Important Exam Idea</u>

For T2, remember this simple rule: chlorine is stronger, and chloramine lasts longer.

Chloramines are usually used for residual protection in the distribution system, not as the main high-speed primary disinfectant.

Common Operator Concerns

Chloramination can be affected by source water ammonia, temperature, pH, storage time, and low water use periods. Warm water and stagnant conditions make nitrification more likely, so operators watch these systems closely.

Taste and odor problems can happen if chloramines are not controlled properly. That is why good monitoring of ammonia, chlorine residual, and system conditions is important.

Simple Memory Trick

Chlorine plus ammonia equals a long-lasting residual, but operators must watch for nitrification.

CHLORINE DIOXIDE

Chlorine dioxide is a strong oxidant and disinfectant used in some drinking water systems. It is handled differently from chlorine and chloramines because it does not mainly work as a persistent residual disinfectant. When used in water, it reacts rapidly and forms chlorite as its main byproduct. Chlorate is also possible.

Chlorine dioxide is a reactive disinfectant gas dissolved in water that can kill microorganisms effectively. It is not the same as chlorine, and it behaves more like a selective oxidant than a typical chlorination chemical.

Water systems may use chlorine dioxide for taste and odor control, for primary disinfection, or when they want a disinfectant that works differently from free chlorine. It can be useful when operators want to reduce certain chlorine-related byproducts.

The major concern with chlorine dioxide is the formation of chlorite and chlorate. These are regulated because they can pose health concerns, and they are the main things operators monitor when chlorine dioxide is used.

EPA guidance commonly cited for drinking water is 0.8 mg/L for chlorine dioxide residual and 1.0 mg/L for chlorite. For exam purposes, remember that chlorine dioxide has a tight residual limit because the byproducts are the real issue.

Operator Concerns

The biggest operational issue is controlling the dose so the system gets disinfection without overproducing chlorite or chlorate. Because chlorine dioxide breaks down quickly, operators focus more on dose control and byproduct control than on leaving a long-lasting distribution residual.

DBPS

Disinfection byproducts, or DBPs, are chemicals that can form when disinfectants react with natural organic matter or other compounds in water. The big ones for T2 are TTHMs, HAA5, chlorite, chlorate, and sometimes bromate, depending on the disinfectant used.

DBPs matter because operators need disinfection that is strong enough to protect public health, but not so aggressive that it creates excessive byproducts. Operators often have to balance microbial safety with chemical compliance.

What Forms DBPs

Chlorine can form THMs and HAAs.

Chloramines usually form fewer THMs than free chlorine, but they can cause nitrification problems.

Chlorine dioxide mainly forms chlorite and chlorate. Ozone can form bromate if bromide is present.

LOG REMOVAL

Log removal is a way to measure how much a treatment process reduces microorganisms. A higher log number means greater removal or inactivation.

Basic Meaning

- 1-log = 90% removed.
- 2-log = 99% removed.
- 3-log = 99.9% removed.
- 4-log = 99.99% removed.

For surface water systems, treatment goals often include 3-log Giardia and 4-log virus removal or inactivation. That means the plant must combine filtration and disinfection to reach the required level of protection.

DBPs and log removal are connected because stronger disinfection can improve microbial safety, but it can also increase some byproducts. So for T2, think of it as a tradeoff: more disinfection often means more attention to DBPs.

EXAM MEMORY

DBP question = think chlorine, chloramines, ozone, chlorine dioxide, bromate, THMs, and HAA5.

Log question = think 1-log = 90%, 2-log = 99%, 3-log = 99.9%, and 4-log = 99.99%.

What Each Log Is For

- 4-log = viruses.
- 3-log = Giardia.
- 2-log = Cryptosporidium.
- 1-log = extra general reduction credit, depending on the treatment process.

CHLORINE CYLINDERS

Chlorine is commonly stored as a compressed liquefied gas in steel cylinders or larger containers. Small cylinders are typically the 100- and 150-lb cylinders, while larger systems may use ton containers, tank cars, or cargo tanks. Because chlorine is toxic and corrosive, even a small leak can become an emergency.

Safety Basics

Chlorine cylinders must be handled by trained personnel using proper PPE and respiratory protection. Operators should keep cylinders secured upright, protect them from heat and damage, and make sure valves, piping, and connections are in good condition. Good ventilation is important because chlorine gas can quickly create a hazardous atmosphere.

Preventing Leaks

Most chlorine leaks are prevented by good handling and inspection rather than repairs. Key prevention steps include:

- Check the cylinder and valve area for damage before use.
- Keep valve caps in place when not connected.

- Use the correct pigtails, washers, and gaskets.
- Open valves slowly and carefully.
- Store cylinders where they will not be struck, corroded, or overheated.

Common Leak Points

Leaks usually occur at the valve, fusible plug, or side wall of the cylinder. On larger containers, leaks can also occur at pressure relief devices or angle valves. That is why emergency kits are built for specific container types.

The Main Chlorine Emergency Kits

Kit A is for 100- and 150-lb chlorine cylinders. It is used to stop leaks at the cylinder valve, fusible plug, and side wall. This is the main kit for small chlorine cylinders.

Kit B is for one-ton chlorine containers. It is used to contain leaks around the ton container valves and side wall. This is the main kit for ton containers.

Kit C is for chlorine tank cars, cargo tanks, and portable tanks. It is designed for leaks in and around the pressure relief device and angle valves. This is the large-container kit.

Simple Memory Trick

- A = cylinder.
- B = ton container.
- C = car, truck, or tank.

What to Remember for the Exam

The kits are made to contain leaks, not to replace proper emergency response. Trained personnel with the right respiratory protection must handle chlorine leaks.

MAIN TYPES OF MICROORGANISMS IN WATER

The main types of microorganisms in water are bacteria, viruses, protozoa, and sometimes algae.

- Bacteria are single-celled organisms. Examples include coliforms, E. coli, Salmonella, Legionella, and Pseudomonas.
- Viruses are the smallest infectious microorganisms. Examples include enteroviruses, norovirus, rotavirus, hepatitis A, and adenovirus.
- Protozoa are single-celled parasites, such as Giardia and Cryptosporidium.
- Helminths, or parasites, are less common in treated drinking water, but they are part of the broader microbial contaminant group.
- Algae and cyanobacteria are more of a source-water and taste-and-odor issue than a classic drinking-water pathogen issue, but they are still important in treatment.

Sampling

Sampling means collecting water correctly so lab results are accurate for bacteria, chemicals, and process control. It matters because a bad sample can give a false result and lead to the wrong operational decision.

Main Sample Types

- Microbiological samples: total coliform, E. coli, Giardia, Cryptosporidium, and other microbes.
- Chemical samples: chlorine residual, nitrate, nitrite, metals, and other water quality parameters. • Process samples: samples used by operators to check treatment performance, such as turbidity or disinfectant residual.

MICROBIOLOGICAL SAMPLING

For bacteriological sampling, the bottle must be sterile and usually contains sodium thiosulfate if the water is chlorinated. This prevents the disinfectant from continuing to kill bacteria after collection. Operators should not rinse the bottle and should avoid touching the inside of the cap or bottle.

The tap should be prepared properly, and in many cases, the water is allowed to run before collecting the sample.

Chemical samples are collected differently depending on the test. Some samples need a clean bottle, some need preservation, and some need cooling right away. The exact sampling method depends on whether the test is for chlorine, nutrients, metals, or other constituents.

IMPORTANT SAMPLING RULES

- Use the correct bottle from the lab.
- Do not rinse the bottle unless the lab says to.
- Avoid contaminating the sample.
- Label the sample correctly and record the location, date, and time.
- Keep samples at the proper temperature and deliver them on time.

Why Sampling Matters in EXAM

Sampling usually connects to coliform monitoring, chlorine residual, and compliance sampling. The key idea is that accurate sampling is just as important as the test itself.

MAIN TYPES OF SAMPLING IN WATER TREATMENT

The main types of sampling in water treatment are grab sampling and composite sampling.

Grab Sampling

A grab sample is a single sample taken at one place and one time. It gives a snapshot of water quality at that exact moment.

Use grab samples for things that can change quickly, such as:

- pH.
- Chlorine residual.
- Dissolved gases.
- Instantaneous contamination events.

Composite Sampling

A composite sample is made by combining multiple small samples collected over time. It shows the average water quality over a period, such as 8, 12, or 24 hours.

Composite sampling can be:

- Time proportional, using equal volumes at equal time intervals.
- Flow proportional, where the sample amount changes with flow.

Manual vs. Automatic Sampling

- Manual sampling means the operator collects the sample by hand.
- Automatic sampling means a sampler collects samples on a schedule or based on flow.

When to Use Each

Use grab sampling when you need a moment-in-time result.

Use composite sampling when you want an average over time.

EXAM Memory Trick

- Grab = now.
- Composite = over time.

POSITIVE TOTAL COLIFORM OR E. COLI SAMPLE

If a sample is positive for total coliform or E. coli, the operator should act immediately and follow the rule-required response steps.

First Actions

Do not ignore it. A positive result can indicate contamination, a treatment problem, or a distribution system breach.

Collect the required repeat samples within 24 hours for total coliform positives.

If the lab or rule requires it, have the sample analyzed for E. coli.

Do not shock chlorinate before collecting repeat samples unless the regulator approves it because that can affect the investigation and follow-up results.

If E. coli Is Found

An E. coli-positive result is more serious and usually requires immediate notification and additional response steps. In groundwater systems, triggered source sampling and corrective actions may be required depending on the rule and result.

Simple Memory Trick

Positive sample means repeat sample, investigate, and do not shock chlorinate first.

CHAIN OF CUSTODY

Chain of custody is the documented record that shows who collected, handled, transferred, stored, and analyzed a water sample from start to finish. It is important because drinking-water data may be used for compliance or legal purposes, so the sample has to be traceable and defensible.

A chain-of-custody form usually records the sample ID, location, date and time, collector, preservatives used, requested analysis, and every transfer of possession. The goal is to prove the sample was not tampered with or mixed up.

It is especially important for compliance samples, investigative samples, and any sample that could be used in legal or regulatory action. If the chain is broken, the sample data can be challenged.

<u>A Sample Is in Custody If It Is</u>

- In someone's physical possession.
- In their view after possession. • Locked or secured so only authorized people can access it.

CHAPTER 5 REVISION

What is the main purpose of disinfection in drinking water treatment?

A. Remove hardness
B. Kill or inactivate disease-causing organisms
C. Raise pH
D. Add minerals

Answer: B

Which disinfectant is most associated with a measurable distribution-system residual?

A. UV
B. Ozone
C. Chlorine
D. Boiling

Answer: C

What does a higher CT generally mean?

A. Less disinfection
B. More disinfection
C. More turbidity
D. Lower pH only

Answer: B

What is a grab sample?

A. A sample collected over 24 hours
B. A single sample taken at one time and place
C. A filtered sample
D. A sample with preservative only

Answer: B

What is a composite sample?

A. One sample only
B. A sample mixed from multiple collections over time
C. A sample from one tap only
D. A sterile sample only

Answer: B

Which sample type gives a snapshot of water quality?

A. Grab

B. Composite

C. Automatic only

D. Duplicate

Answer: A

Which sample type gives an average over time?

A. Grab

B. Composite

C. Field blank

D. Split sample

Answer: B

What is the chain-of-custody document?

A. Chlorine dose

B. Sample handling history

C. Filter run time

D. Tank level

Answer: B

Why is chain of custody important?

A. It improves taste

B. It proves the sample was handled properly

C. It adds residual

D. It reduces turbidity

Answer: B

What should an operator do after a positive total coliform sample?

A. Ignore it

B. Collect repeat samples promptly

C. Turn off pumps

D. Add lime

Answer: B

Why are positive E. coli results serious?

A. They mean pH is low
B. They indicate potential fecal contamination
C. They always mean chlorine is too high
D. They are harmless

Answer: B

What are the main microorganism groups important in drinking water?

A. Rocks, metals, plastics
B. Bacteria, viruses, and protozoa
C. Sand, silt, and clay
D. Odor, taste, and color

Answer: B

Which protozoan is a common drinking-water concern?

A. Giardia
B. Iron bacteria
C. Algae
D. Yeast

Answer: A

Which organism is generally more resistant to chlorine?

A. Protozoa
B. Table salt
C. Oxygen
D. Sand

Answer: A

What does 1-log removal equal?

A. 10%
B. 50%
C. 90%
D. 99.9%

Answer: C

What does 2-log removal equal?

A. 90%

B. 99%

C. 99.9%

D. 10%

Answer: B

What does 3-log removal equal?

A. 99.9%

B. 99%

C. 90%

D. 99.99%

Answer: A

What does 4-log removal equal?

A. 99%

B. 99.9%

C. 99.99%

D. 90%

Answer: C

In the 4-3-2 rule, what log level is usually associated with viruses?

A. 1-log

B. 2-log

C. 3-log

D. 4-log

Answer: D

In the 4-3-2 rule, what log level is usually associated with Giardia?

A. 4-log

B. 3-log

C. 2-log

D. 1-log

Answer: B

In the 4-3-2 rule, what log level is usually associated with Cryptosporidium?

A. 4-log

B. 3-log

C. 2-log

D. 1-log

Answer: C

What is a common formula for free chlorine residual?

A. Dose minus demand

B. Demand minus dose

C. pH minus alkalinity

D. Flow times time

Answer: A

Adding too much lime

A. Reaching the point where added chlorine satisfies demand and free chlorine appears

B. Turning off a filter

C. UV failure

Answer: B

What is chloramine formed from?

A. Chlorine and ammonia

B. Chlorine and lime

C. Ozone and bromide

D. UV and iron

Answer: A

Which chloramine species is usually desired in drinking water?

A. Trichloramine

B. Dichloramine

C. Monochloramine

D. Hypochlorite

Answer: C

Why do utilities use chloramines?

A. Shorter residual

B. Longer-lasting residual

C. Higher turbidity

D. No residual needed

Answer: B

What is a major problem in chloraminated systems?

A. Freezing

B. Nitrification

C. Sedimentation

D. Fluoridation

Answer: B

What can nitrification cause?

A. Higher chloramine residual

B. Loss of residual and nitrite formation

C. More hardness

D. More pressure

Answer: B

Which disinfectant inactivates organisms without leaving a residual?

A. Chlorine

B. Chloramine

C. UV

D. Lime

Answer: C

Which disinfectant is strong and can reduce taste-and-odor compounds?

A. Ozone

B. Sand

C. Fluoride

D. Sulfur

Answer: A

Which disinfectant mainly forms chlorite as a byproduct?

A. Chlorine dioxide

B. UV

C. Lime

D. Chloramine

Answer: A

What is the main concern with chlorine dioxide?

A. Iron staining

B. Chlorite and chlorate byproducts

C. Hardness

D. Alkalinity loss

Answer: B

What is a common feature of UV disinfection?

A. It adds residual

B. It is a chemical disinfectant

C. It inactivates organisms without leaving residual

D. It creates THMs

Answer: C

What is ozone generated with?

A. A chlorine cylinder

B. An ozone generator

C. A filter press

D. A lime slurry tank

Answer: B

What is a common drawback of ozone?

A. No effect on microbes

B. No lasting residual

C. Always high pH

D. Increases hardness

Answer: B

DBPs?

A. Disinfection byproducts

B. Daily balance points

C. Distribution bulk pumps

D. Dissolved bacteria particles

Answer: A

Which are common chlorine DBPs?

A. THMs and HAAs

B. Nitrate and fluoride

C. Iron and manganese

D. Chloride and sodium

Answer: A

What DBP is associated with chlorine dioxide?

A. Bromate

B. Chlorite

C. Fluoride

D. Sulfate

Answer: B

What DBP can ozone form if bromide is present?

A. Bromate

B. Chlorite

C. Chlorate

D. Nitrate

Answer: A

Why do operators monitor pH in disinfection?

A. pH affects chlorine effectiveness

B. pH changes color only

C. pH controls flow

D. pH removes metals

Answer: A

Chlorine generally works best under what pH condition?

A. Higher pH

B. Lower pH

C. Neutral only

D. pH does not matter

Answer: B

What affects chlorine demand?

A. Organic matter and other reducing substances

B. Pipe color only

C. Tank shape only

D. Pump size only

Answer: A

What is a clearwell mainly used for?

A. Filter media storage

B. Contact time for disinfection

C. Sludge drying

D. Chemical mixing only

Answer: B

Why is baffling important?

A. It shortens contact time

B. It improves effective contact time

C. It raises hardness

D. It adds chlorine demand

Answer: B

What does poor baffling do?

A. Improves CT

B. Reduces CT efficiency

C. Removes DBPs

D. Increases filter media size

Answer: B

What is a chlorine cylinder commonly made of?

A. Plastic

B. Steel

C. Glass

D. Copper

Answer: B

What is one key safety rule for chlorine cylinders?

A. Store in sunlight

B. Secure upright and protect from damage

C. Open quickly

D. Use without ventilation

Answer: B

What should be done if a chlorine leak is suspected?

A. Ignore it

B. Evacuate and isolate the area

C. Add more chlorine

D. Rinse the floor only

Answer: B

What is Kit A used for?

A. Tank cars

B. One-ton containers

C. 100- and 150-lb chlorine cylinders

D. UV units

Answer: C

What is Kit B used for?

A. Chlorine cylinders

B. One-ton chlorine containers

C. Pump seals

D. Filter media

Answer: B

What is Kit C used for?

A. Small cylinders

B. Tank cars and large tanks

C. Chemical day tanks

D. Sample bottles

Answer: B

What is a common chlorine leak point?

A. Valve or fusible plug

B. Filter underdrain

C. Turbidity meter

D. pH probe

Answer: A

What does the chain-of-custody form record?

A. Chemical feed rate only

B. Who handled the sample and when

C. Filter backwash frequency

D. Tank pressure only

Answer: B

Why should sterile bottles not be rinsed for bacteriological samples?

A. To add preservatives

B. To avoid contamination and preserve sample integrity

C. To increase pH

D. To remove chlorine residual

Answer: B

Why may sodium thiosulfate be in a bacteriological sample bottle?

A. To increase hardness

B. To neutralize disinfectant residual

C. To raise pH

D. To remove turbidity

Answer: B

What is a positive total coliform result an indicator of?

A. Distribution system or sample contamination issue

B. High hardness

C. Low temperature only

D. Excess fluoride

Answer: A

What is the first follow-up for a positive coliform sample?

A. Repeat sampling

B. Shut the plant forever

C. Add lime

D. Lower turbidity with chlorine

Answer: A

Why is E. coli especially important?

A. It is harmless

B. It suggests fecal contamination

C. It indicates high alkalinity

D. It is a disinfectant

Answer: B

What is a common purpose of sampling in water treatment?

A. To decorate reports

B. To verify water quality and compliance

C. To replace disinfection

D. To raise chlorine demand

Answer: B

What are the main disinfectants commonly discussed in the exam preparation ?

A. Chlorine, chloramines, chlorine dioxide, ozone, and UV

B. Sand, gravel, clay, silt, and rock

C. Lime, soda ash, caustic, alum, and ferric

D. Carbon, sulfur, iron, manganese, and nitrate

Answer: A

Which disinfectant is both a disinfectant and oxidant but not a long-residual disinfectant?

A. Chlorine dioxide
B. Chloramine
C. Lime
D. Fluoride

Answer: A

What does log removal describe?

A. Chemical concentration in mg/L
B. Pathogen reduction in powers of ten
C. Filter head loss
D. Pressure in psi

Answer: B

What does reducing 10,000 organisms to 10 organisms represent?

A. 1-log
B. 2-log
C. 3-log
D. 4-log

Answer: C

What does reducing 10,000 organisms to 1 organism represent?

A. 1-log
B. 2-log
C. 3-log
D. 4-log

Answer: D

Which process physically removes particles and some microbes before disinfection?

A. Filtration
B. Fluoridation
C. Corrosion control
D. Softening only

Answer: A

What is one reason operators monitor chlorine residual in the distribution system?

A. To know if disinfectant protection remains

B. To raise hardness

C. To add turbidity

D. To remove alkalinity

Answer: A

What is a common response to nitrification in chloraminated water?

A. Stop all pumping

B. Flush, reestablish residual, and control ammonia

C. Add more algae

D. Remove all chlorine

Answer: B

Which is a physical disinfection method?

A. UV

B. Chlorine

C. Chloramine

D. Chlorine dioxide

Answer: A

What is the best overall exam memory for chlorine safety?

A. Chlorine is harmless if small

B. Chlorine leaks require training, PPE, and emergency response kits

C. Keep cylinders open

D. Use no ventilation

Answer: B

CHAPTER 6

PUMPS, VALVES, AND MOTORS

Pump Course Topics

Pump course topics include:

- Pump basics: what pumps do, head, flow, pressure, and horsepower.
- Pump types: centrifugal pumps and positive displacement pumps.
- Pump curves: how to read flow, head, efficiency, horsepower, and NPSH.
- Cavitation and NPSH: why suction conditions matter.
- System curves: how the system affects pump performance.
- Operation and maintenance: priming, valves, seals, alignment, vibration, and troubleshooting.

COMMON EXAM-STYLE PUMP QUESTIONS

For exam-style study, the most common pump questions are about:

- Head vs. pressure.
- Flow vs. pressure.
- Pump curve interpretation.
- NPSH and cavitation.
- Centrifugal vs. positive displacement pumps.
- Priming.
- Valves.
- Seal leaks.
- Troubleshooting low flow or loss of prime.

Pump Types to Learn

Centrifugal pumps are common in water treatment. Their flow changes with head and pressure. Positive displacement pumps deliver nearly constant flow and are better for viscous fluids or chemical feed.

Submersible pumps are often used in wells or lift stations.

Vertical turbine pumps are often used for wells and high-lift service.

Easy Memory Trick

- Centrifugal = changing flow.
- Positive displacement = nearly constant flow.
- Low suction pressure = cavitation risk.

What Pumps Do

Pumps move water from one place to another by adding energy to the water. In water treatment, pumps are used for raw water, finished water, chemical feed, and backwash systems.

Head is the height a pump can raise water, usually measured in feet of head. It is closely related to pressure. For water, 2.31 feet of head equals 1 psi.

Head is often used instead of pressure because it is easier to compare pump performance across different fluids.

Flow is how much water the pump moves in a given time, usually measured in gallons per minute, or gpm. A pump curve shows how flow changes as head changes. In general, for centrifugal pumps, as flow goes up, head goes down.

Pressure is the force the water exerts per unit area, usually measured in psi. In pump work, pressure and head are related, but pump operators usually talk about head because it is more useful for comparing performance. Higher pressure usually means the pump is working against more resistance.

Horsepower is the power needed to move water. Water horsepower can be estimated with this formula:

Water horsepower = (Flow in gpm × Head in feet) ÷ 3960

If efficiency is included, the pump needs more power than water horsepower alone.

Easy Example

If a pump moves 100 gpm against 50 feet of head, then water horsepower is:

Water horsepower = (100 × 50) ÷ 3960 = 1.26 hp

That means the pump needs about 1.26 water horsepower before efficiency losses are considered.

Memory Trick

- Flow = how much water.
- Head = how high or how hard the pump pushes.
- Pressure = force on the water.
- Horsepower = power required to do the work.

CENTRIFUGAL PUMPS

Centrifugal pumps use a spinning impeller to add energy to the fluid and move it through the pump. They are the most common pumps in water treatment because they work well with low-viscosity liquids like water and can move high flow rates efficiently.

Their flow changes as system pressure changes, so they do not deliver exactly the same flow at every operating point.

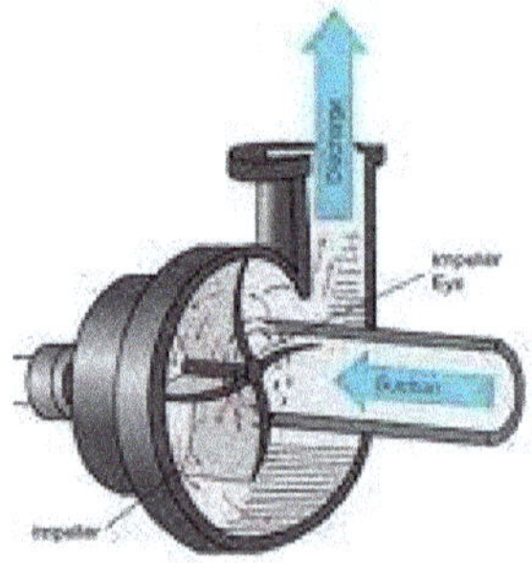

POSITIVE DISPLACEMENT PUMPS

Positive displacement pumps trap a fixed volume of liquid and force it out into the discharge line. They are best for thick, viscous, or metered fluids, such as chemical feed or sludge-type applications.

They give a more constant flow even when pressure changes.

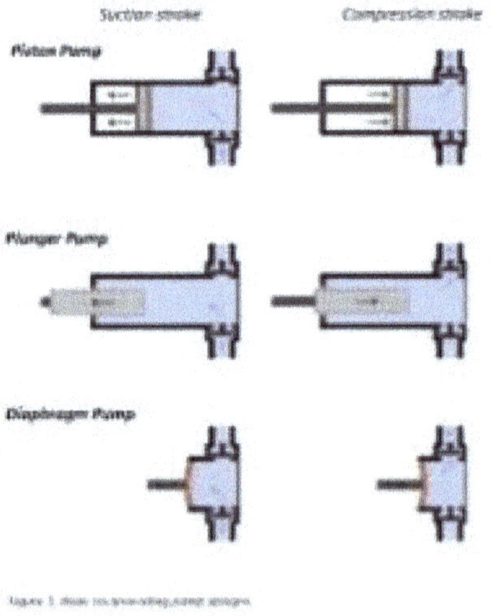

- [ ] remember
- Centrifugal: spinningimpeller,changingflow

Easy Memory Trick

- Positive displacement = trapping volume, steady flow.

CENTRIFUGAL PUMP TYPES

Main Subtypes

- End-suction pumps are the most common type, with fluid entering one side of the impeller.
- Split-case pumps are high-flow pumps with a casing split
- Vertical inline pumps are compact pumps with suction and discharge in line.
- Vertical turbine pumps are used for deep wells, reservoirs, and raw water intakes.
- Multistage pumps use multiple impellers to produce higher head.
- Submersible pumps operate underwater and are common in sump

Easy Memory Trick

- End-suction = everyday use.
- Split-case = big flow.
- Vertical turbine = deep sources.
- Multistage = high head.
- Submersible = underwater service.

POSITIVE DISPLACEMENT PUMP TYPES

Positive displacement pumps are usually grouped into reciprocating and rotary types.

Reciprocating Pumps

Reciprocating pumps use back-and-forth motion to move liquid.

Common types include:

- Piston pumps.
- Plunger pumps.
- Diaphragm pumps.

Rotary Pumps

Rotary pumps use rotating parts to trap and move liquid.

Common types include:

- Gear pumps.
- Vane pumps.
- Lobe pumps.
- Screw pumps.
- Progressive cavity pumps.
- Peristaltic pumps.

Easy Memory Trick

- Reciprocating = back and forth.
- Rotary = spinning.

MAIN PARTS OF A CENTRIFUGAL PUMP

Here are the main parts of a centrifugal pump you should know for T2:

- Impeller: the rotating part that adds energy to the water.
- Casing: the outer housing that collects and directs flow.
- Shaft: connects the motor to the impeller.
- Bearings: support the shaft and reduce friction.
- Seal or packing: prevents leakage along the shaft.
- Wear rings: reduce internal leakage and protect pump parts.

Other Common Parts

- Shaft sleeve: protects the shaft from wear and corrosion.
- Coupling: connects the pump and motor.
- Gaskets and O-rings: help prevent leakage at joints.
- Volute or diffuser: helps convert velocity into pressure.

Easy Memory Trick

Think of a centrifugal pump like this:

- Impeller = moves water.
- Casing = collects water.
- Shaft = spins water mover.
- Bearings = support the shaft.
- Seal = stops leaks.
- Wear rings = reduce internal leakage.

ROUTINE CHECKS

Operators should regularly inspect for leaks, unusual noise, vibration, and overheating. They should also check suction and discharge pressure because changes can show a developing problem.

Bearings must be kept properly lubricated with the correct oil or grease. Poor lubrication can lead to bearing wear, higher temperatures, and failure.

Mechanical seals should be inspected for leakage, and packing should stay properly lubricated. If packing becomes dry or compressed, it may need replacement.

Check coupling alignment between the pump and motor because misalignment causes vibration and wear. Excessive vibration can also point to cavitation, bearing problems, or blockage. During shutdown or overhaul, inspect the impeller, casing, wear rings, and shaft for erosion, corrosion, cracks, or buildup. A damaged impeller can reduce flow and efficiency. Keep suction strainers, screens, and lines clean so the pump does not starve for water. A restricted suction line can contribute to cavitation and poor performance.

PUMP MAINTENANCE

Pump maintenance includes seals, bearings, alignment, lubrication, vibration checks, and inspection.

Common Maintenance Schedule

- Daily: check leaks, noise, vibration, and temperatures.
- Weekly or monthly: check lubrication, alignment, and performance.
- Annual: inspect and rebuild internal parts as needed.

COMMON CENTRIFUGAL PUMP FAILURES

The most common problems are loss of prime, low flow, cavitation, vibration, seal leaks, overheating, and bearing failure. These often show up as noise, poor pressure, unstable flow, or excess heat.

If the pump runs but delivers little or no flow, it may have lost prime. Common causes are air leaks on the suction side, a clogged foot valve or strainer, or low source water level. Check suction piping, fittings, and valve condition first.

Low flow can happen when the impeller is worn or clogged, suction is restricted, speed is too low, or the system demand is too high. Operators should check suction strainers, discharge restrictions, impeller condition, and whether the pump is operating near its best efficiency point. Cavitation happens when pressure inside the pump drops too low and vapor bubbles form and collapse. It often sounds like gravel or rattling and can damage the impeller, casing, and shaft. Fix it by improving suction conditions, cleaning strainers, reducing suction lift, or lowering speed if needed.

Vibration is often caused by cavitation, misalignment, worn bearings, impeller imbalance, or pipe strain. Check alignment, bearing condition, and whether the impeller is damaged or loaded with buildup. Vibration should never be ignored because it often leads to bigger failures.

Leaks around the shaft usually point to a worn mechanical seal, packing problem, or dry running.

If the seal area is leaking, inspect the seal face, packing adjustment, and flush conditions.

Continued leakage can lead to bearing damage and overheating.

Overheating usually means bearing problems, poor lubrication, dry running, or too little flow. Check the oil or grease level, bearing condition, and whether the pump is running closed or nearly closed. Heat is a warning sign, not just a nuisance.

CAVITATION IN THE FIELD

In the field, look first for the classic signs: gravel-like noise, vibration, reduced discharge pressure or flow, and possible seal or bearing damage. In centrifugal pumps, cavitation usually shows up as noise plus lower pressure and poor performance.

Field Signs

- Sound: rattling, crackling, popping, or marbles/gravel noise.
- Vibration: higher-than-normal vibration or shaking in the pump and piping.
- Performance drop: reduced flow or reduced discharge pressure.
- Damage clues: impeller pitting or erosion, seal leaks, or bearing failure if cavitation has been happening for a while.

QUICK FIELD CHECKS

Check the suction side first for clogged strainers, partially closed valves, low tank level, or air leaks.

Compare suction pressure and discharge pressure to normal operating values.

Look at whether the pump is operating too far from its best efficiency point.

If allowed, a simple test is to slowly throttle the discharge valve. If noise and vibration improve, cavitation is more likely.

Why It Happens

Cavitation usually means the pump does not have enough NPSH available at the suction side, so vapor bubbles form and collapse near the impeller.

Common causes include suction restrictions, poor piping, high suction lift, low liquid level, or operating too far left on the pump curve.

Simple Memory Trick

Noise, vibration, low pressure, and poor flow mean cavitation should be suspected.

CHEMICAL PUMPS

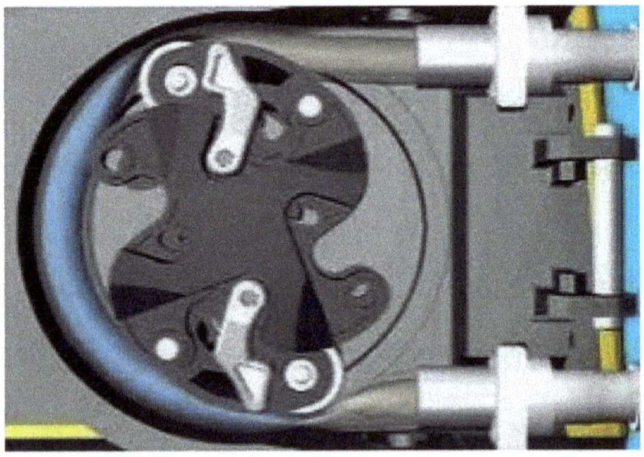

Chemical pumps are usually chemical metering pumps or chemical feed pumps that inject an exact amount of chemical into a water stream. In water treatment, they are used for chlorine, sodium hypochlorite, acids, caustic, corrosion control chemicals, and other treatment chemicals. Chemical pumps dose chemicals accurately and consistently instead of just moving large volumes of water. They are used when the plant needs controlled addition, such as disinfection, pH adjustment, coagulation aid feed, or corrosion control.

COMMON CHEMICAL FEED PUMPS

The most common chemical feed pumps in treatment plants are:

- Diaphragm metering pumps.
- Peristaltic pumps.
- Plunger or piston metering pumps.
- Electromagnetic metering pumps for smaller or more precise applications.

Diaphragm metering pumps are popular because they are positive displacement pumps and can handle aggressive chemicals with good control. They use a flexible diaphragm and check valves to draw chemical in and then push it out in a measured stroke.

<u>Common Uses</u>

Chemical pumps are used for:

- Chlorine or hypochlorite feed.
- Acid or caustic feed for pH control.
- Coagulant or polymer feed.
- Corrosion control and other process chemicals.

<u>Key Exam Ideas</u>

For T2, remember:

- Chemical pumps are usually metering pumps.
- They are often positive displacement pumps.
- They are chosen for accuracy and repeatability.
- Calibration matters because the actual feed rate must match the intended dose.

<u>Easy Memory Trick</u>

Think: chemical pump = precision feed pump.

ROUTINE MAINTENANCE

Routine maintenance usually includes checking for leaks, cleaning filters, and inspecting the pump and piping for wear. Operators should also inspect the injection quill and make sure it is not clogged. Chemical storage tanks should be kept clean so sludge or solids do not plug the system.

Chemical feed pumps should be calibrated regularly to confirm they are delivering the correct chemical dose. Calibration is especially important after installation, repair, or any major process change. Many systems use a calibration cylinder to measure actual pump output.

Maintenance Checks

A good maintenance check includes:

- Pump output compared with setpoint.
- Suction line, foot valve, and screen condition.
- Injection quill condition and clogging.
- Leaks in tubing, fittings, and valves.
- Pressure gauge readings.
- Chemical tank level and cleanliness.

Operators may need to flush the pump, tubing, and injection point to keep chemical buildup from hardening and restricting flow. This is especially important for chemicals that can crystallize or leave solids.

Pump components wear out over time, so parts like diaphragms, check valves, seals, and tubing may need replacement. If output changes, pressure fluctuates, or leaks appear, worn parts are often the cause.

Chemical pump maintenance should always include safe handling of the chemical being fed. Operators should isolate the pump, relieve pressure, and use proper PPE before servicing.

Simple Memory Trick

Inspect, clean, calibrate, and replace worn parts.

PUMP CURVE

☐ pumpcurve

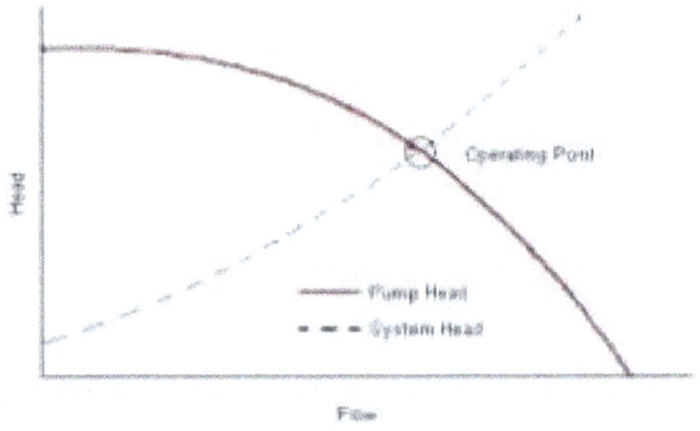

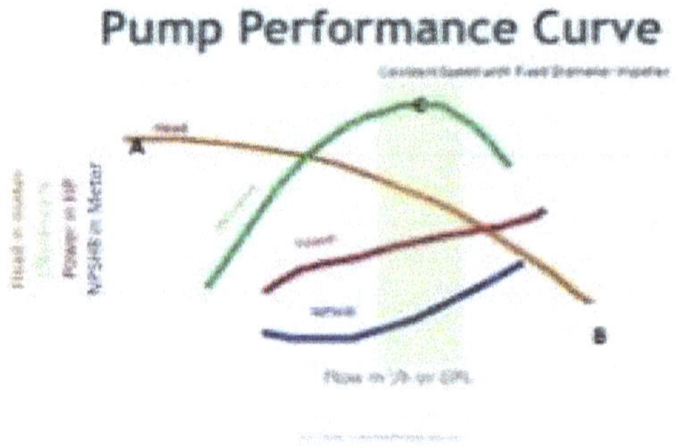

On most pump curves, flow is on the x-axis and head is on the y-axis. Flow is how many gallons per minute the pump delivers, and head is the energy the pump adds to the water, often shown in feet.

The efficiency curve shows how well the pump converts input power into useful water movement. The highest point on the efficiency curve is the best efficiency point, or BEP. Pumps usually operate most smoothly and economically near BEP.

The horsepower curve shows how much power the pump needs at different flow rates. As flow and head change, the required horsepower changes too. If the pump is pushed to higher head or flow, the horsepower demand usually rises.

NPSH

The NPSH curve shows how much suction head the pump requires to avoid cavitation. NPSHr is the required suction head shown on the pump curve, and NPSHa is what the system can actually provide. For safe operation, NPSHa must be greater than NPSHr.

<u>How to Use the Curve</u>

1. Find the required flow along the bottom axis.
2. Move up to the head curve to see the pump's head at that flow.
3. Read the efficiency curve to see how close the pump is to BEP.
4. Check the horsepower curve to make sure the motor is large enough.
5. Check NPSHr so the pump will not cavitate.

<u>Memory Trick</u>

- Flow = how much water.
- Head = how hard the pump pushes.
- Efficiency = how well it works.
- Horsepower = how much power it needs.
- NPSH = suction safety margin.

HOW TO CALCULATE NPSH IN THE FIELD

To calculate NPSH available in the field, use the suction-side conditions at the pump and compare them to the pump's required NPSH. For a simple field calculation on a centrifugal pump, the common form is:

NPSHa = absolute pressure head + static head - suction-line friction loss - vapor pressure

<u>What You Usually Need</u>

- Suction pressure converted to absolute pressure.
- Liquid temperature to find vapor pressure.
- Static suction lift or flooded suction.
- Suction piping friction loss.

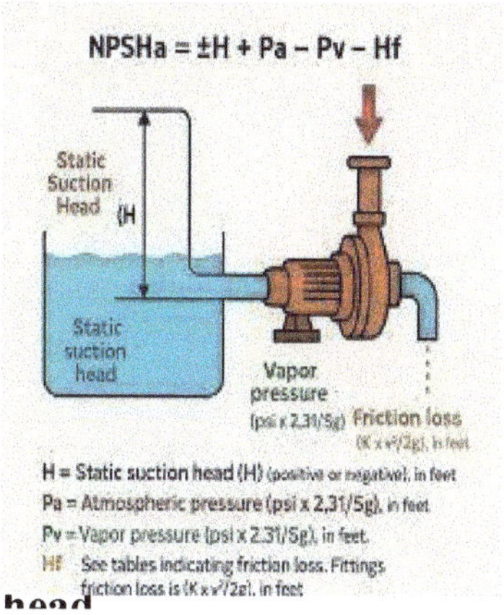

head

<u>Simple Field Steps</u>

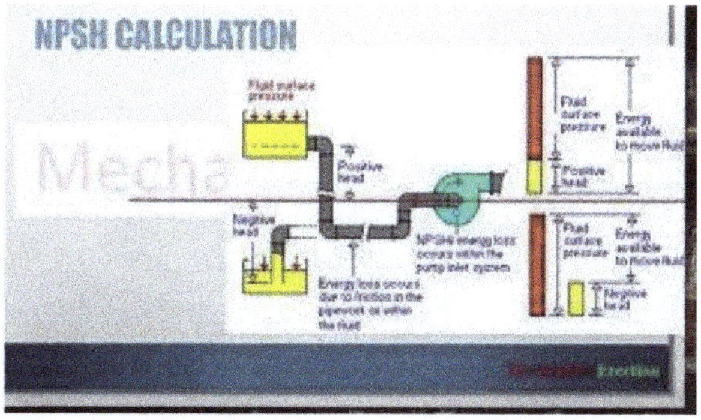

Measure suction pressure at the pump and convert it to absolute pressure if needed.

Convert that pressure to feet of head.

Determine whether the pump has flooded suction or suction lift. Add static head if the pump has flooded suction, and subtract static head if the pump has suction lift.

Estimate suction-line friction losses from pipe, valves, fittings, and strainers.

Subtract vapor pressure head based on the liquid temperature.

Compare NPSHa to NPSHr from the pump curve. NPSHa must be greater than NPSHr.

Quick Example

If the suction system gives:

- Absolute pressure head = 34 ft.
- Static head = 10 ft.
- Friction loss = 3.2 ft.
- Vapor pressure head = 0.8 ft. Then:

NPSHa = 34 + 10 - 3.2 - 0.8 = 40 ft

That means the pump has about 40 feet of available suction head.

MOTORS

What a Motor Does

An electric motor converts electrical energy into mechanical rotation that drives a pump, blower, or other machine. In water treatment, motors are essential because they provide the torque and power needed to move water and chemicals.

Main Motor Types

- AC motors: the most common in water treatment.
- Single-phase motors: used for smaller loads and simpler applications.
- Three-phase motors: common in municipal and industrial pump service because they are efficient and provide strong starting and running performance.
- Induction motors: the most common motor type in pumping applications.
- DC motors: less common in typical plant pumping service.

Common Motor Parts

- Stator: the stationary part that creates the magnetic field.
- Rotor: the rotating part that turns the shaft.
- Shaft: transfers mechanical power to the pump.
- Bearings: support the rotor and reduce friction.
- Enclosure: protects the motor from dust, moisture, and contamination.

Motor Maintenance

Good motor maintenance usually includes:

- Checking for noise, heat, smell, and vibration.
- Inspecting bearings and lubrication condition.
- Keeping the motor clean and dry.
- Checking alignment between the motor and pump.
- Inspecting wiring, terminals, and insulation condition.

Common Field Problems

Common field problems include:

- Overheating from overload, poor ventilation, or bearing problems.
- Vibration from misalignment or worn bearings.
- Tripping or failure to start from electrical issues or a bad starting component.
- Noise from bad bearings or mechanical binding.

Exam Memory Tricks

- Single-phase = smaller/simple.
- Three-phase = plant standard.
- TEFC = most protected/common in treatment plants.
- Motor problems often show up as heat, vibration, noise, or trips.
- Know the difference between TEFC and ODP.
- Remember basic maintenance and troubleshooting.

VALVES

MAIN TOPICS

Valve topics include:

- Valve purpose: start, stop, isolate, regulate, and prevent backflow.
- Common valve types: gate, globe, check, ball, butterfly, plug, diaphragm, pinch, and needle valves.
- How valves operate: quarter-turn, multi-turn, linear motion, and self-actuated valves.
- Applications in water treatment: isolation, throttling, backflow prevention, and flow control.
- Parts of valves: body, bonnet, stem, disc, seat, packing, actuator, and handwheel.
- Maintenance: leaks, packing adjustment, corrosion, exercising valves, and lubrication where needed.
- Troubleshooting: stuck valves, leaking valves, hard-to-turn valves, and failed check valves.

EASY MEMORY TRICK

- Gate = isolate.
- Globe = regulate.
- Ball = quick on/off.
- Butterfly = large pipe.
- Check = no backflow.
- Plug = rotary.
- Diaphragm = corrosive.
- Pinch = slurry.
- Needle = precision.

COMMON VALVE TYPES

Gate valves are best for start/stop service and isolation.

Globe valves are good for regulating or throttling flow.

Check valves prevent reverse flow automatically.

Backflow preventers protect potable water from contamination by stopping reverse flow.

GATE VALVE

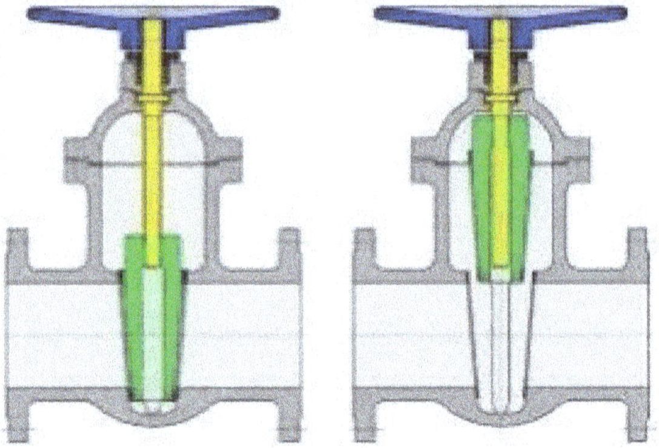

A gate valve course for T2 should focus on what gate valves do, how they work, where they are used, and how to maintain them.

MAIN TOPICS

- Purpose: gate valves are mainly used for start/stop and isolation service.
- Operation: the gate moves up and down to open or block flow.
- Common uses: water supply, wastewater, fire mains, and plant isolation.
- Advantages: low pressure drop when fully open and good shutoff for isolation.
- Parts: body, bonnet, stem, gate/disc, seat, packing, and handwheel.
- Maintenance: lubrication, leak checks, corrosion inspection, exercising the valve, and cleaning sediment buildup.

What a Gate Valve Is Used For

A gate valve is mainly used for start/stop service and isolation of a pipe, pump, tank, or section of a system. When it is fully open, it gives a straight flow path with low resistance.

How It Works Mechanically

A gate valve works by moving a gate or wedge up and down inside the valve body. The handwheel turns the stem, which raises or lowers the gate. Fully raised means open, and fully lowered means closed.

Why It Is Not Usually Used for Throttling

A gate valve is not usually used for throttling because partial opening creates turbulence, vibration, erosion, and possible cavitation, which can damage the seat and gate.

Gate valves are designed for fully open or fully closed service, not for precise flow control.

How to Inspect and Maintain It

Inspect a gate valve for stem leaks, packing leaks, corrosion, debris buildup, and smooth operation. Maintenance usually includes exercising the valve, checking packing and stem condition, and making sure the valve opens and closes without binding.

How to Recognize Common Failures

Common failures include a valve that is stuck, hard to turn, leaking at the stem or seat, or not fully opening or closing. These problems are often caused by corrosion, worn packing, damaged seats, or sediment and debris inside the valve.

Exam Memory Trick

Think: gate valve = isolate, not throttle.

When fully open, it should give minimal resistance.

GLOBE VALVE

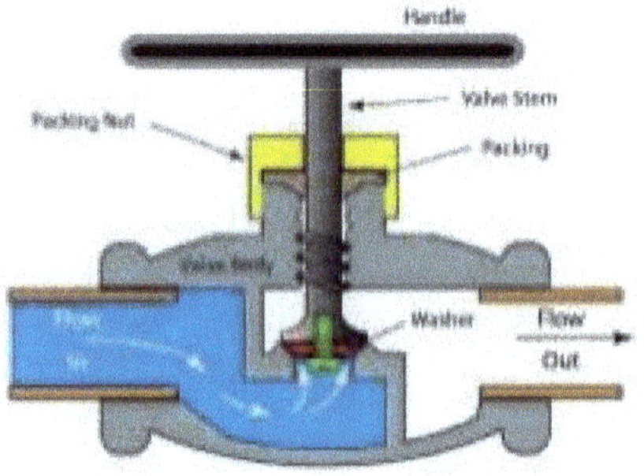

A globe valve is mainly used for throttling and flow regulation. It can also be used to start and stop flow, but its best strength is controlling how much water passes through.

A globe valve uses a movable disc or plug that moves up and down against a seat. As the stem turns, the opening gets larger or smaller, which lets the operator adjust flow more precisely than with a gate valve.

Why It Is Good for Throttling

Globe valves are designed for fine control, so they handle partial opening better than gate valves. Their flow path creates more resistance, but that same design makes them useful when operators need to regulate pressure or flow steadily.

In practice, they are often the best textbook choice for throttling service.

Maintenance and Inspection

Inspect globe valves for packing leaks, seat wear, corrosion, and stem stiffness. Maintenance usually includes checking the handwheel or actuator, exercising the valve, and making sure the disc seals properly.

If the valve is hard to operate or leaks when closed, the seat or packing may need repair.

Typical failures include leakage, internal seat wear, corrosion, and difficulty opening or closing. Because globe valves are often used in throttling service, worn trim and erosion can happen over time if flow is very fast or dirty.

Common Applications

Globe valves are mainly used for throttling and flow control in water treatment. They are also used where an operator needs fine adjustment of flow or pressure rather than just full open/full closed service.

Common applications include:

- Chemical dosing and feed lines: used to regulate chemical flow precisely.
- Raw water and finished water systems: used to control flow and pressure in treatment and distribution lines.
- Pumping stations: used for flow balancing and pressure control.
- Filtration systems: used to adjust flow through filters and treatment units.
- Sludge and waste lines: used where controlled flow is needed.

Globe valves are good for applications that need gradual adjustment because their design gives better control than a gate valve. That makes them useful in places where process conditions need to stay stable.

<u>Exam Memory Trick</u>

Think: globe valve = control valve.

If the question is about precise throttling, globe valve is usually the right answer.

BALL VALVE

A ball valve is a quarter-turn valve used for quick start/stop control of flow. In water treatment, it is often used for isolation, shutoff, and some chemical or water line service.

A ball valve has a hollow ball with a hole through it. When the handle turns 90 degrees, the hole lines up with the pipe and flow passes through. When turned again, the solid part of the ball blocks flow.

Ball valves are popular because they are fast, simple, and seal well. They are good for frequent on/off operation and can be useful where quick isolation is needed.

COMMON USES IN WATER TREATMENT

- Isolation valves for pumps, tanks, and piping sections.
- Chemical feed lines and other clean-fluid services.
- Emergency shutoff when fast response is needed.

Ball valves are common in water treatment because they provide fast shutoff, tight sealing, and simple isolation.

MAINTENANCE AND INSPECTION

Check ball valves for external leaks, stem issues, seal wear, and debris buildup. Operators should cycle valves periodically so they do not seize, and clean them when needed.

If a ball valve starts leaking internally or becomes hard to turn, the seats or seals may be worn.

COMMON APPLICATIONS

- Isolation of pumps, tanks, and piping sections during maintenance.
- Chemical feed and dosing lines where fast shutoff is useful.
- Raw water, filtration, and finished water lines for on/off control.
- Emergency shutoff and backflow protection support in safety systems. Why They Are Used

Ball valves are popular because they have quarter-turn operation, low pressure drop when fully open, and bubble-tight shutoff. That makes them practical for quick isolation and reliable service in plant operations.

EXAM MEMORY TRICK

Think: ball valve = quick isolation valve.

It is usually better for on/off service than for fine throttling.

- Ball valve = quarter-turn.
- Best for quick shutoff.
- Good seal and low maintenance.

BUTTERFLY VALVE

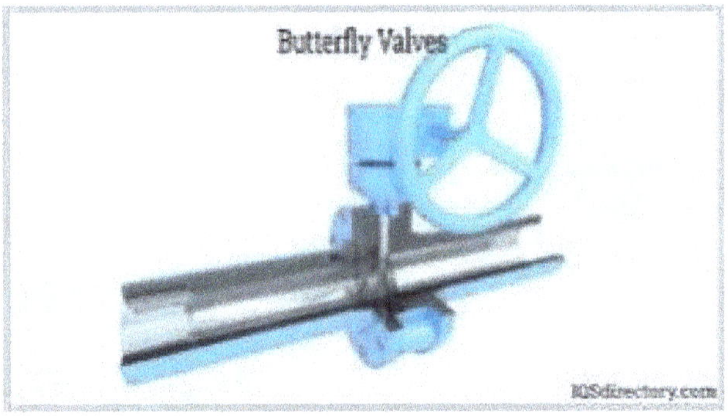

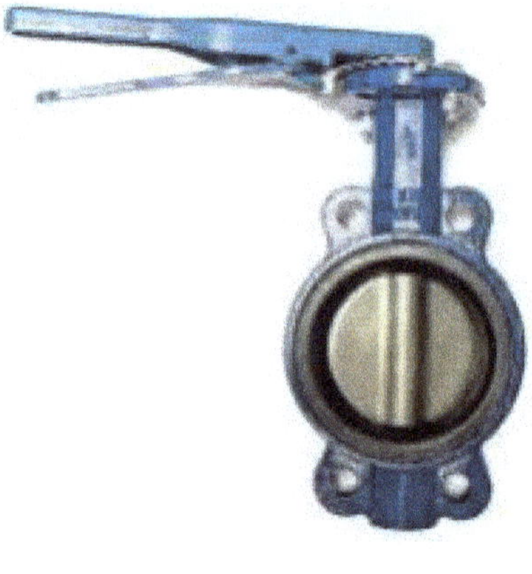

A butterfly valve is a quarter-turn valve used to start, stop, isolate, and sometimes regulate flow in water treatment systems. It uses a rotating disc that turns 90 degrees to open or close the line.

<u>How It Works</u>

When the disc is parallel to flow, the valve is open. When the disc is perpendicular to flow, the valve is closed. The valve body, stem, seat, and disc work together to control the fluid path.

COMMON WATER TREATMENT USES

- Water supply and distribution lines.
- Pumping stations.
- Isolation service for large pipelines and tanks.
- Flow regulation where moderate control is acceptable.

Why Operators Like Them

Butterfly valves are compact, lightweight, and cost-effective. They are often used in large-diameter piping because they take up less space than many other valve types.

Maintenance and Inspection

Check the disc, seat, stem, bearings, and actuator for wear or corrosion. Routine maintenance includes cycling the valve, cleaning debris, and making sure the seat still seals properly. If the valve is not used often, it should be exercised regularly so it does not seize.

Common Failures

Common problems include seat wear, leakage, stiff movement, and disc damage. If the valve is used too aggressively for throttling, the seat or disc can be damaged.

Exam Memory Trick

- Butterfly = quarter-turn.
- Good for large pipes.
- Often used for isolation and moderate control.

Think: butterfly valve = large pipe, quick isolation, moderate control.

They are useful, but they are not the best choice when very precise throttling is needed over long periods.

CHECK VALVES

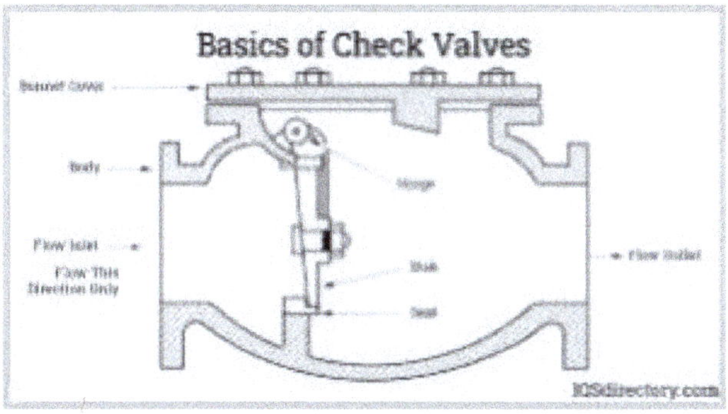

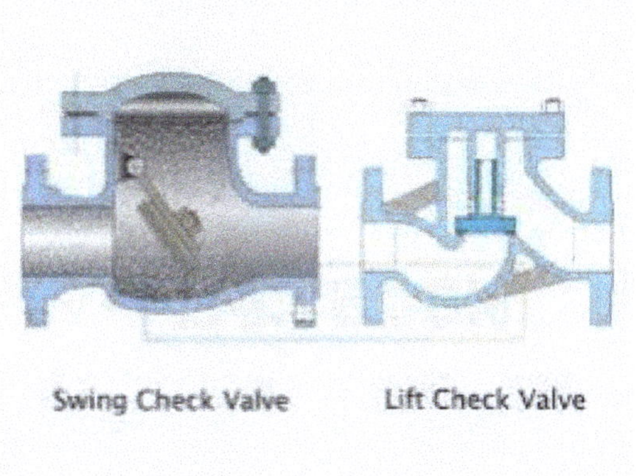

Check valves are used to prevent backflow in water and wastewater systems.

A check valve allows flow in one direction only and closes automatically if flow reverses. This protects pumps, piping, tanks, and meters from reverse flow damage.

Common Types

Swing check valves are the most common type in water distribution and pumping service.

Lift check valves are used where fluid flow lifts the disc and reverse flow closes it.

Other styles may include ball, dual-plate, and spring-assisted designs.

WHERE THEY ARE USED

Check valves are commonly used in:

- Pumping stations.
- Lift stations.
- Water distribution lines.
- Wastewater force mains.
- Tank and reservoir connections.

Why They Matter

They stop reverse flow, help protect pumps from having to re-pump the same water, and reduce damage from backflow and water hammer.

In simple terms, they keep the system moving in the right direction.

Maintenance and Inspection

Check valves usually need regular inspection for leaks, corrosion, debris, and smooth operation. A common check is to confirm the valve moves freely and seats properly when flow stops or reverses.

If the valve is noisy, leaking, or not closing, it may need cleaning, repair, or replacement. T2 Memory Trick

Think: check valve = one-way valve.

If flow tries to go backward, the check valve closes automatically.

PLUG VALVE

A plug valve is used for quick on/off control, isolation, and flow diversion in water treatment systems. In water and wastewater plants, the most common style is often the eccentric plug valve because it handles dirty water and solids better and gives tight shutoff.

COMMON APPLICATIONS IN WATER TREATMENT

- Wastewater and sewage service: collection, separation, aeration, and discharge.
- Sludge and solids-heavy lines: good where grit or solids could clog other valves.
- Water and sewage treatment plants: used for isolation and shutoff.
- Flow diversion: useful in lines that need to switch flow paths, such as backwash lines.

Why They Are Useful

Plug valves are valued for simple design, low operating torque, reliable shutoff, and durability. Eccentric plug valves are especially good in wastewater because they can seal well even with solids present.

Exam Memory Trick

Think: plug valve = strong shutoff for dirty service.

They are often chosen when the water is not perfectly clean and the valve must still close tightly.

DIAPHRAGM VALVE

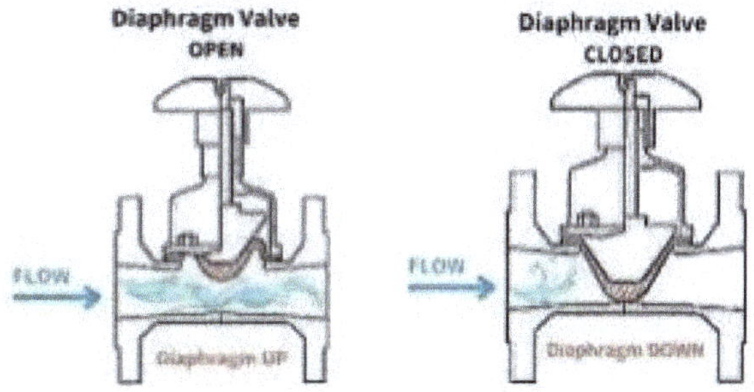

A diaphragm valve is used when good shutoff and protection from corrosive or dirty fluids are needed. It works by pressing a flexible diaphragm against a seat, which keeps the process fluid away from the stem and moving parts.

In water treatment, diaphragm valves are common in chemical dosing lines, corrosive chemical handling, wastewater service, and slurry or sludge service. They are also used in sanitary or high-purity service because the flow path is simple and easy to clean.

The main advantage is that the diaphragm isolates the working parts from the fluid, which reduces leakage and corrosion problems. That makes diaphragm valves a strong choice for acids, bases, and other aggressive chemicals.

Common issues include diaphragm wear, leakage, residue buildup, and stem or seat damage. Regular inspection should focus on the diaphragm condition, seat sealing, and any sign of cracking, swelling, or chemical attack.

T2 Memory Trick

Think: diaphragm valve = corrosive and dirty service.

If the fluid is harsh or the valve needs a very clean sealing barrier, diaphragm valves are often the right answer.

CHAPTER 6 PRACTICE EXAMS

What is the main purpose of a centrifugal pump?

A. Store water

B. Move water by adding energy

C. Remove solids from water

D. Measure flow

Answer: B

Which part of a centrifugal pump rotates?

A. Casing

B. Impeller

C. Suction pipe

D. Volute

Answer: B

Which pump part surrounds the impeller?

A. Motor

B. Casing

C. Seal ring

D. Bearing

Answer: B

What is the function of the pump shaft?

A. Direct flow

B. Convert chemical energy

C. Transfer rotation to the impeller

D. Prevent cavitation

Answer: C

What do bearings do in a pump?

A. Increase pressure

B. Support the shaft and reduce friction

C. Stop backflow

D. Measure head

Answer: B

What is the purpose of a mechanical seal?

A. Increase flow

B. Prevent leakage along the shaft

C. Reduce motor speed

D. Clean the impeller

Answer: B

What is packing used for?

A. To lubricate the motor

B. To seal around the shaft

C. To stop the impeller from spinning

D. To measure suction pressure

Answer: B

Which pump component provides a pressure boundary?

A. Shaft

B. Bearing

C. Casing

D. Coupling

Answer: C

Which part of the pump adds energy to the water?

A. Impeller

B. Seal

C. Bearing

D. Gasket

Answer: A

What is the main function of a pump?

A. Convert electricity to chemistry

B. Move liquid

C. Remove chlorine

D. Increase temperature

Answer: B

What is cavitation?

A. A clogged discharge line

B. Vapor bubbles forming and collapsing in a pump

C. A broken seal

D. A type of motor

Answer: B

What is a common sign of cavitation?

A. Smooth, silent operation

B. Gravel-like noise

C. No motor current

D. No suction line

Answer: B

What usually causes cavitation?

A. Too much lubrication

B. Not enough suction head

C. Too large a pipe

D. Low voltage only

Answer: B

What does NPSH stand for?

A. Net Pump Speed Head

B. Net Positive Suction Head

C. Normal Pressure Suction Heat

D. Net Primary System Head

Answer: B

Why is NPSH important?

A. It tells motor size

B. It helps prevent cavitation

C. It measures efficiency only

D. It controls chemical feed

Answer: B

What does the casing do in a centrifugal pump?

A. Rotates the shaft

B. Holds the liquid and directs flow

C. Lubricates bearings

D. Measures pressure

Answer: B

What is the suction side of a pump?

A. The side where liquid enters

B. The side where liquid leaves

C. The motor side only

D. The seal side only

Answer: A

What is the discharge side of a pump?

A. Where liquid enters

B. Where liquid leaves

C. Where the motor starts

D. Where air is released

Answer: B

Which pump part is connected to the motor and impeller?

A. Shaft

B. Volute

C. Casing

D. Packing

Answer: A

What is a wear ring used for?

A. Increase suction lift

B. Reduce internal leakage and wear

C. Start the motor

D. Measure flow rate

Answer: B

What is the best reason to inspect pump seals?

A. To increase pipe diameter

B. To stop leaks

C. To change speed

D. To reduce voltage

Answer: B

Which problem often causes pump overheating?

A. Good alignment

B. Bearing failure

C. Clean suction line

D. Correct NPSH

Answer: B

What is one main job of a bearing?

A. Hold the flow rate steady

B. Support rotating parts

C. Close the discharge valve

D. Remove air from the casing

Answer: B

Why do pumps need proper lubrication?

A. To increase pH

B. To reduce friction and wear

C. To make flow turbulent

D. To seal the discharge line

Answer: B

What happens if a pump loses prime?

A. It may stop moving liquid

B. It becomes a valve

C. It increases head

D. It removes cavitation

Answer: A

What is a common cause of loss of prime?

A. Full suction line

B. Air entering the suction side

C. Excess bearing grease

D. High efficiency

Answer: B

What should be checked first if a pump has low flow?

A. Pump color

B. Suction restrictions

C. Paint condition

D. Motor label only

Answer: B

What does pump efficiency tell you?

A. How much power is wasted

B. How well the pump converts input power into flow

C. The pipe size

D. The valve opening

Answer: B

Which part keeps the process liquid from leaking along the rotating shaft?

A. Bearing

B. Seal

C. Casing drain

D. Coupling guard

Answer: B

Which of the following is a rotating pump part?

A. Casing

B. Impeller

C. Packing gland

D. Volute

Answer: B

What does a pump curve show?

A. Chemical dosage only

B. Flow, head, horsepower, efficiency, and NPSH

C. Pipe color and size

D. Valve type only

Answer: B

On a pump curve, what is on the x-axis?

A. Head

B. Flow

C. Horsepower

D. Speed only

Answer: B

What does head represent?

A. Water color

B. Energy added to the liquid

C. Pump age

D. Pipe length only

Answer: B

What is the best efficiency point called?

A. NPSH

B. BEP

C. RPM

D. CAV

Answer: B

Why is BEP important?

A. It is where the pump is most efficient

B. It is where the pump stops

C. It is where the valve closes

D. It is where flow becomes zero

Answer: A

What does horsepower on a pump curve tell you?

A. Motor color

B. Power required by the pump

C. Pipe friction only

D. Chemical demand

Answer: B

What does NPSHr mean?

A. Net Pump Speed Rating

B. Net Positive Suction Head Required

C. Normal Pressure System Height Required

D. Net Pump Suction Handling Rate

Answer: B

What does NPSHa mean?

A. Net Positive Suction Head Available

B. Normal Pump Speed Actual

C. Net Pressure Safety Height Available

D. New Pump Suction Head Area

Answer: A

What must be true to avoid cavitation?

A. NPSHa must be less than NPSHr

B. NPSHa must equal zero

C. NPSHa must be greater than NPSHr

D. NPSHr must be negative

Answer: C

cavitation

A. A valve type

B. Vapor bubbles forming and collapsing in the pump

C. A bearing lubricant

D. A flow meter reading

Answer: B

common sign of cavitation is:

A. Smooth operation

B. Gravel-like noise

C. Lower pipe diameter

D. Higher pH

Answer: B

What often causes cavitation?

A. Too much suction head

B. Too little NPSH available

C. Too many bearings

D. Too much pipe slope

Answer: B

Which condition can help prevent cavitation?

A. Increase suction restrictions

B. Lower liquid level

C. Reduce suction friction losses

D. Close the discharge valve fully

Answer: C

What can happen if cavitation continues?

A. Pump parts can be damaged

B. Pump becomes more efficient

C. Water disappears

D. Valve alignment improves

Answer: A

What part of the pump is often damaged by cavitation?

A. Impeller

B. Motor nameplate

C. Handwheel

D. Pipe coating only

Answer: A

What is a common cause of low pump flow?

A. Dirty suction strainer

B. Perfect alignment

C. Good lubrication

D. High NPSHa

Answer: A

What should be checked first if pump flow is low?

A. Suction side restrictions

B. Paint quality

C. Valve handle color

D. Bearing brand

Answer: A

What is a likely cause of pump vibration?

A. Cavitation

B. Good suction piping

C. Proper flow

D. Low noise

Answer: A

What is another cause of vibration?

A. Misalignment

B. Clean strainer

C. Proper lubrication

D. Correct speed

Answer: A

What is the purpose of pump maintenance?

A. Make it heavier

B. Prevent failure and keep it operating properly

C. Increase pipe corrosion

D. Reduce pressure to zero

Answer: B

Which pump part should be checked for leakage?

A. Seal

B. Bearing cap only

C. Motor fan only

D. Pipe elbow only

Answer: A

What does a bearing support?

A. The rotating shaft

B. The chemical tank

C. The valve stem only

D. The suction tank

Answer: A

Why are bearings important?

A. They increase cavitation

B. They reduce friction and support rotating parts

C. They close the pump discharge

D. They control flow rate

Answer: B

What can happen if lubrication is poor?

A. Lower wear

B. Overheating and bearing failure

C. Better sealing

D. More suction head

Answer: B

What is pump priming?

A. Painting the pump

B. Filling the pump and suction line with liquid

C. Removing the impeller

D. Opening the motor starter

Answer: B

Why must a pump be primed?

A. To prevent it from running dry

B. To increase pipe friction

C. To close the discharge valve

D. To reduce motor size

Answer: A

What happens if a centrifugal pump runs dry?

A. It usually improves

B. It can overheat and be damaged

C. It becomes self-priming

D. It automatically stops forever

Answer: B

What is the discharge side of a pump?

A. The suction entrance

B. The outlet where liquid leaves

C. The motor cooling side

D. The seal water inlet only

Answer: B

What is one sign of a worn mechanical seal?

A. No leakage

B. Leakage around the shaft

C. Higher NPSH

D. Lower horsepower only

Answer: B

What is the main purpose of a valve?

A. Add oxygen to water

B. Control, start, or stop flow

C. Increase motor speed

D. Remove solids

Answer: B

Which valve is best for fully open or fully closed service?

A. Globe valve

B. Gate valve

C. Needle valve

D. Check valve

Answer: B

Which valve is commonly used for throttling and flow control?

A. Globe valve

B. Check valve

C. Gate valve

D. Plug valve

Answer: A

Which valve is used to prevent backflow automatically?

A. Ball valve

B. Check valve

C. Gate valve

D. Butterfly valve

Answer: B

Which valve uses a rotating disc?

A. Gate valve

B. Butterfly valve

C. Globe valve

D. Needle valve

Answer: B

Which valve gives a straight-through flow path?

A. Ball valve

B. Globe valve

C. Check valve

D. Diaphragm valve

Answer: A

Which valve is often used for isolation in large-diameter pipelines?

A. Butterfly valve

B. Needle valve

C. Check valve

D. Stop valve only

Answer: A

Which valve can be used for precise flow regulation but causes more pressure drop?

A. Globe valve

B. Gate valve

C. Ball valve

D. Check valve

Answer: A

What does a check valve stop?

A. High voltage

B. Backflow

C. Pump cavitation

D. Motor overheating

Answer: B

What is a quarter-turn valve?

A. A valve turned 90 degrees to open or close

B. A valve turned once every quarter hour

C. A valve that needs four turns

D. A valve used only on quarter-inch pipe

Answer: A

Which valve is often used in sludge service because it can handle thick material?

A. Knife gate valve

B. Needle valve

C. Check valve

D. Globe valve

Answer: A

Which valve is best known for on/off service and good shutoff?

A. Ball valve

B. Globe valve

C. Check valve

D. Air release valve

Answer: A

What is throttling?

A. Turning a pump off

B. Regulating flow rate

C. Increasing pipe size

D. Cleaning the valve seat

Answer: B

Which valve is least suitable for throttling over long periods?

A. Butterfly valve

B. Globe valve

C. Needle valve

D. Control valve

Answer: A

What is a common use of butterfly valves in wastewater plants?

A. Pump lubrication

B. Flow isolation and regulation

C. Chemical analysis

D. Air compressor cooling

Answer: B

What does a gate valve usually do best?

A. Fine control

B. Full open or full close

C. Prevent all pressure loss

D. Measure flow

Answer: B

Which valve is designed to allow flow in one direction only?

A. Check valve

B. Gate valve

C. Globe valve

D. Ball valve

Answer: A

Which valve type is generally compact and cost-effective for large pipes?

A. Butterfly valve

B. Globe valve

C. Needle valve

D. Plug valve

Answer: A

Which valve is commonly used where accurate throttling is needed?

A. Globe valve

B. Gate valve

C. Check valve

D. Butterfly valve

Answer: A

What is the main purpose of a ball valve?

A. Backflow prevention only

B. Quick shutoff with low pressure drop

C. Precise throttling only

D. Pump priming

Answer: B

Which valve is most likely to be used at a pump discharge to prevent reverse flow?

A. Check valve

B. Needle valve

C. Gate valve

D. Globe valve

Answer: A

What happens when a butterfly valve disc turns parallel to flow?

A. Flow stops

B. Flow passes freely

C. Pressure doubles

D. Cavitation begins

Answer: B

What happens when a butterfly valve disc turns perpendicular to flow?

A. Flow stops

B. Flow increases

C. Pump speed rises

D. Water level drops

Answer: A

Which valve is best for very fine flow adjustment?

A. Needle valve

B. Gate valve

C. Check valve

D. Butterfly valve

Answer: A

Which valve automatically responds to flow direction?

A. Check valve

B. Ball valve

C. Gate valve

D. Globe valve

Answer: A

What is a major advantage of a gate valve?

A. Great for throttling

B. Low head loss when fully open

C. Automatic operation

D. Prevents all cavitation

Answer: B

What is a major disadvantage of a globe valve?

A. No flow control

B. Higher pressure drop

C. Cannot be opened

D. Too fast to operate

Answer: B

Which valve is often used in chemical feed lines for isolation?

A. Ball valve

B. Check valve

C. Air valve

D. Knife gate valve

Answer: A

Which valve is generally not used to stop reverse flow by itself?

A. Check valve

B. Gate valve

C. Ball valve

D. Butterfly valve

Answer: B

Which of the following is a valve used for isolation?

A. Gate valve

B. Pump seal

C. Bearing

D. Impeller

Answer: A

What is a motor used for in a pump system?

A. To store water

B. To provide rotating power

C. To measure pressure

D. To stop backflow

Answer: B

Which electrical problem can cause a motor to overheat?

A. High voltage only
B. Low voltage or overload
C. Clean bearings
D. Open discharge valve

Answer: B

What is one sign of motor trouble?

A. Smooth operation
B. Excessive heat
C. Low friction
D. Clear water

Answer: B

What should be checked if a pump will not start?

A. Valve color
B. Power supply and controls
C. Pipe coating
D. Water temperature only

Answer: B

What is a coupling used for?

A. Joining the motor shaft to the pump shaft
B. Stopping leakage
C. Measuring flow
D. Opening a valve

Answer: A

Why is alignment important?

A. It reduces vibration and wear
B. It increases cavitation
C. It lowers flow to zero
D. It closes the motor

Answer: A

What can happen if a pump is misaligned?

A. Less vibration

B. More vibration and bearing wear

C. Better efficiency

D. Higher water quality

Answer: B

What is a common reason for a pump losing flow?

A. Clogged strainer

B. Correct speed

C. Good lubrication

D. Open suction line

Answer: A

What should be inspected when a pump has abnormal noise?

A. Seal, bearings, and cavitation signs

B. Wall color

C. Pipe age only

D. Motor label only

Answer: A

What is the purpose of a strainer?

A. Increase pressure

B. Remove debris from the flow

C. Add chlorine

D. Stop the motor

Answer: B

Which valve is used to prevent reverse flow?

A. Check Valve

B. Gate Valve

C. Globe Valve

D. Needle valve

Answer: A

Which valve is best for full open/full closed service?

A. Gate valve

B. Globe valve

C. Needle valve

D. Control valve

Answer: A

Which valve is commonly used for throttling?

A. Globe valve

B. Gate valve

C. Check valve

D. Plug valve

Answer: A

Which valve uses a rotating disc?

A. Butterfly valve

B. Gate valve

C. Check valve

D. Diaphragm valve

Answer: A

Which valve is usually opened with a quarter turn?

A. Ball valve

B. Globe valve

C. Gate valve

D. Needle valve

Answer: A

What is the main advantage of a ball valve?

A. Precise throttling

B. Quick shutoff

C. High pressure drop

D. Automatic backflow prevention

Answer: B

Which valve is compact and common on larger pipelines?

A. Butterfly valve

B. Needle valve

C. Globe valve

D. Check valve

Answer: A

Which valve is best for very fine adjustment?

A. Needle valve

B. Gate valve

C. Check valve

D. Butterfly valve

Answer: A

Which valve is often used in sludge service?

A. Knife gate valve

B. Needle valve

C. Check valve

D. Globe valve

Answer: A

What does throttling mean?

A. Turning off the pump

B. Regulating flow rate

C. Increasing pipe size

D. Removing air

Answer: B

Which valve usually causes more pressure loss?

A. Globe valve

B. Gate valve

C. Ball valve

D. Check valve

Answer: A

Which valve is not usually used for fine throttling?

A. Butterfly valve

B. Globe valve

C. Needle valve

D. Control valve

Answer: A

What does a gate valve do best?

A. Fine flow control

B. Full open or full close

C. Automatic backflow stop

D. Pressure measurement

Answer: B

Which valve can be used to isolate equipment?

A. Gate valve

B. Pump seal

C. Bearing

D. Impeller

Answer: A

What is a check valve used for?

A. To prevent reverse flow

B. To increase suction head

C. To reduce motor size

D. To remove solids

Answer: A

What happens when a butterfly valve disc is parallel to flow?

A. Flow is open

B. Flow is blocked

C. Pressure drops to zero

D. Pump stops

Answer: A

What happens when a butterfly valve disc is perpendicular to flow?

A. Flow stops

B. Flow increases

C. Head rises

D. Cavitation stops

Answer: A

Which valve is best for accurate throttling?

A. Globe valve

B. Gate valve

C. Ball valve

D. Check valve

Answer: A

Which valve has a straight-through flow path?

A. Ball valve

B. Globe valve

C. Needle valve

D. Check valve

Answer: A

Which valve type is generally least suitable for long-term throttling?

A. Butterfly valve

B. Globe valve

C. Needle valve

D. Control valve

Answer: A

What is the first thing to check if a pump will not start?

A. Pipe paint

B. Power supply

C. Valve color

D. Water temperature

Answer: B

What is a common cause of low pump discharge pressure?

A. Clogged suction strainer

B. Perfect alignment

C. Clean bearings

D. Open motor panel

Answer: A

What can cause a pump to lose prime?

A. Air entering the suction side

B. Full casing of water

C. Proper lubrication

D. Open discharge line

Answer: A

What is the safest general first step before pump maintenance?

A. Remove the seal

B. Lock out and tag out power

C. Open every valve

D. Increase speed

Answer: B

What does backwashing do?

A. Adds pressure to the filter

B. Reverses flow to clean the filter media

C. Increases chlorine demand

D. Stops the pump

Answer: B

What is a sign that a filter may need backwashing?

A. Low head loss

B. High head loss

C. Higher pH

D. Lower temperature

Answer: B

What is one benefit of regular maintenance?

A. More breakdowns

B. Longer equipment life

C. More leaks

D. Higher cavitation

Answer: B

What should be inspected if a valve leaks around the stem?

A. Stem packing or seal

B. Impeller

C. Motor fan

D. Suction pipe only

Answer: A

What is a common maintenance task on valves?

A. Replace the whole pipe every week

B. Lubricate moving parts if required

C. Increase water temperature

D. Remove the actuator permanently

Answer: B

What is a valve actuator used for?

A. Turning the valve open or closed

B. Measuring chlorine residual

C. Increasing pipe size

D. Removing debris

Answer: A

What does a knife gate valve handle well?

A. Clear water only

B. Thick slurries and solids

C. Air only

D. High voltage

Answer: B

What is one advantage of a butterfly valve?

A. Very high pressure drop

B. Compact design

C. Slow operation only

D. No isolation use

Answer: B

What is one advantage of a ball valve?

A. Quick quarter-turn operation

B. Best for continuous throttling

C. Needs many turns

D. High pressure loss

Answer: A

What is one disadvantage of a globe valve?

A. Very low pressure drop

B. Higher pressure drop

C. No throttling ability

D. Cannot be opened

Answer: B

Which valve is commonly used for backflow prevention at a pump discharge?

A. Check valve

B. Gate valve

C. Needle valve

D. Globe valve

Answer: A

Which valve is most likely to slam shut if flow reverses suddenly?

A. Check valve

B. Gate valve

C. Ball valve

D. Butterfly valve

Answer: A

What is a common symptom of cavitation?

A. Smooth quiet sound

B. Gravel or rattling noise

C. Better efficiency

D. Lower friction only

Answer: B

What should be done to reduce suction losses?

A. Use smaller suction piping

B. Remove suction restrictions

C. Close suction valve partly

D. Add more bends

Answer: B

What does BEP mean?

A. Best Efficiency Point

B. Backflow Emergency Point

C. Base Energy Pressure

D. Bearing End Position

Answer: A

Why should a pump operate near BEP?

A. To increase wear

B. To run most efficiently and reduce stress

C. To cause vibration

D. To stop flow

Answer: B

What is a common cause of pump vibration?

A. Misalignment

B. Good suction conditions

C. Low noise

D. Clean strainers

Answer: A

What can overheating in a motor or bearing indicate?

A. Normal cooling

B. Friction or overload

C. Better flow

D. Lower pressure only

Answer: B

What is the purpose of a coupling guard?

A. Decorate the pump

B. Protect workers from rotating parts

C. Increase flow

D. Stop cavitation

Answer: B

What should be checked if a valve is hard to turn?

A. Packing, corrosion, or lack of lubrication

B. Water color

C. Pipe insulation

D. Chlorine residual only

Answer: A

Which device is approved for high-hazard backflow protection in the cited practice quiz?

A. Double check valve assembly

B. Simple gate valve

C. Needle valve

D. Plug valve

Answer: A

What is the purpose of a backflow prevention device?

A. Increase head

B. Prevent contaminated water from reversing direction

C. Raise motor speed

D. Clean filters

Answer: B

What does CT mean in water treatment?

A. Chlorine Time

B. Concentration times Contact Time

C. Chemical Treatment

D. Control Tank

Answer: B

Why is CT important?

A. It helps determine disinfection effectiveness

B. It measures pipe diameter

C. It controls pump speed

D. It opens valves Answer: A

What disinfectant is commonly emphasized in operator study because it leaves a residual?

A. Chlorine

B. UV light

C. Heat

D. Sand

Answer: A

What is one reason operators monitor chlorine residual?

A. To confirm disinfection is adequate

B. To increase pump wear

C. To stop filtration

D. To reduce pipe size

Answer: A

CHAPTER 7

CORROSION CONTROL, TASTE, AND ODOR CONTROL

Corrosion Control

Corrosion control in T2 water treatment means adjusting the water so it is less likely to dissolve metals from pipes and plumbing, especially lead and copper.

One common method is adding orthophosphate and keeping the pH in the right range so a thin protective coating can form inside the system.

In simple terms, corrosion control is a treatment step that helps protect water quality as it moves through the distribution system.

What Operators Watch

The main things operators watch are:

- pH.
- Alkalinity.
- How aggressive the water is toward metal surfaces.

For this course, the key idea to remember is that corrosion control is about preventing metal from leaching into drinking water and protecting pipes from damage.

Langelier Saturation Index, or LSI

What You Need

To calculate LSI, you need:

- Measured pH.
- Temperature.
- Total dissolved solids, or TDS.
- Calcium hardness as $CaCO_3$.
- Alkalinity as $CaCO_3$.

How to Interpret It

- LSI = 0 means the water is at equilibrium.
- LSI > 0 means the water tends to form scale.
- LSI < 0 means the water tends to be corrosive or aggressive.

Example

If the measured pH is 7.5 and the calculated saturation pH is 7.9, then:

LSI = 7.5 - 7.9 = -0.4

That means the water is slightly corrosive

Water Stability

Water stability depends on the balance of pH, alkalinity, calcium hardness, and temperature.

How Each Factor Affects Stability

pH shows how acidic or basic the water is. Low-pH water is more corrosive, and high-pH water is more likely to form scale.

Alkalinity acts like a buffer that helps keep pH from changing too fast. Low alkalinity means pH can swing more easily.

Calcium hardness is the amount of dissolved calcium in the water. Low calcium hardness can make water aggressive and more likely to dissolve pipe materials, while high calcium hardness can contribute to scale.

Temperature affects how stable the water is. Warmer water generally increases scaling potential, while cooler water can make water more aggressive depending on the balance.

Short Course Answer

Water stability depends on the balance of pH, alkalinity, calcium hardness, and temperature. If pH or alkalinity is too low, water may be corrosive. If pH, calcium, or temperature are too high, water may form scale.

Simple Rule to Remember

- Low pH + low alkalinity = corrosive water.
- High pH + high calcium + warmer water = scale-forming water.

TYPES OF CORROSION

The main types of corrosion are uniform corrosion, pitting corrosion, galvanic corrosion, and tuberculation.

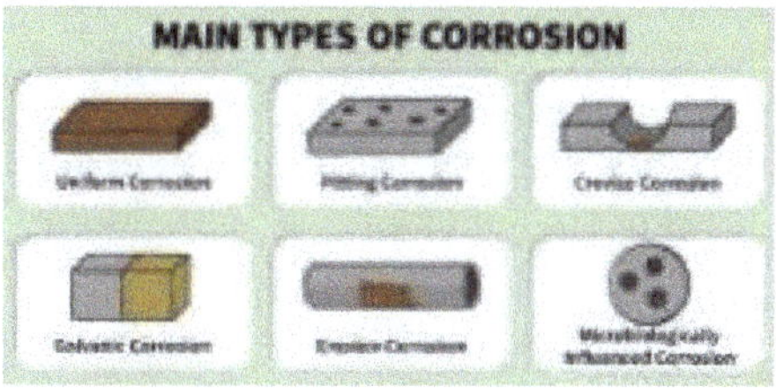

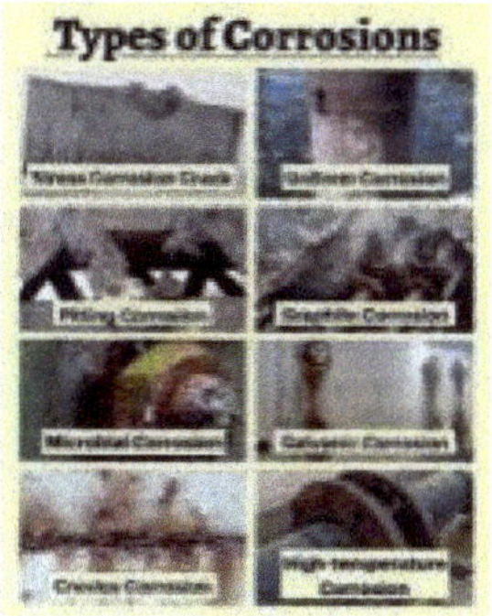

Uniform Corrosion

Uniform corrosion means the metal is eaten away evenly across the whole surface, so the pipe slowly gets thinner.

Pitting Corrosion

Pitting corrosion forms small, deep holes in isolated spots. It is dangerous because the damage can be severe even when most of the surface still looks okay.

Galvanic Corrosion

Galvanic corrosion happens when two different metals touch in water. The less noble metal becomes the anode and corrodes faster.

Tuberculation

Tuberculation happens when corrosion products build up into rough mounds, or tubercles, inside pipes. These deposits often hide underlying metal loss and can restrict flow.

Short Class-Style Answer

Uniform corrosion thins metal evenly, pitting makes small but serious holes, galvanic corrosion occurs between dissimilar metals, and tuberculation is the buildup of corrosion deposits that hide pipe damage.

LEAD AND COPPER RELEASE

Lead and copper can enter drinking water when plumbing materials corrode and dissolve into the water, especially in older pipes, solder, fittings, or service lines. Low-mineral or corrosive water tends to be more aggressive, so it can pull more metal from plumbing.

Key Points

- Lead often comes from old service lines, lead solder, or older brass fixtures.
- Copper usually comes from copper piping, especially when water is corrosive or sits in the pipes too long.
- The longer water stays in contact with plumbing, the more metal can leach into it.
- Hot water can release metals faster than cold water, so cold water is preferred for drinking and cooking.
- Utilities reduce metal release by controlling water chemistry to form protective scale or coating inside pipes.

Simple Class Sentence

Lead and copper release happens when corrosive water reacts with plumbing materials and dissolves metals into drinking water.

HOW CORROSION AFFECTS THE SYSTEM

Corrosion can affect distribution systems, storage tanks, and customer taps. In distribution systems, corrosion can thin pipes, cause leaks, reduce pressure, create tuberculation, and increase the chance of contamination entering through breaks or low-pressure events.

In storage tanks, corrosion can produce rust, metal particles, taste and odor problems, and water quality complaints. It can also increase maintenance and repair costs.

At customer taps, corrosion can release lead, copper, and iron from plumbing, which affects water color, taste, and public health.

SAMPLING, MONITORING, AND TESTING

Sampling, monitoring, and testing are used to see whether corrosion control is working. Utilities commonly collect first-draw tap samples, measure lead and copper, and track water quality parameters such as pH, alkalinity, and related indicators to judge whether the water is stable.

In many programs, the 90th percentile of sampling results is used to evaluate compliance and decide whether additional corrosion control actions are needed.

SAFETY, REGULATIONS, AND PUBLIC HEALTH

Safety, regulations, and public health are central to corrosion control because corrosion can create leaks, structural failures, and contamination risks.

The regulatory goal is to keep lead and copper at safe levels and make sure corrosion control treatment reduces metal release into drinking water.

HOW TO TREAT CORROSION IN A WATER SYSTEM

The main goal is to make the water less aggressive and stop metals from dissolving into pipes and plumbing.

The most common treatments are raising pH and alkalinity, adding corrosion inhibitors like orthophosphate or silicates, and maintaining stable water chemistry so a protective scale or coating can form on pipe walls.

<u>Main Treatment Methods</u>

- Adjust pH and alkalinity with lime, soda ash, or caustic soda to stabilize the water.
- Add orthophosphate to form a protective coating inside lead and copper pipes.
- Use silicates or polyphosphates in some systems to help reduce corrosion and control scale. • Flush problem areas, remove sediment, and rehabilitate or line old pipes when corrosion products have built up.
- Replace highly corroded or incompatible materials when treatment alone is not enough.

<u>What Operators Monitor</u>

After treatment starts, operators usually check pH, orthophosphate residual, alkalinity, lead, copper, iron, and sometimes zinc to see whether corrosion control is working.

The treatment has to be specific to the system because the right dose depends on water quality, pipe materials, and the rest of the treatment process.

<u>Simple Answer</u>

Corrosion is treated by stabilizing the water with pH and alkalinity adjustment, using inhibitors such as orthophosphate, and monitoring the system to make sure metals at the tap stay low.

What is the primary purpose of corrosion control in water distribution systems?

A. Remove bacteria

B. Protect pipes from deterioration and prevent metal leaching

C. Increase flow rate

D. Reduce turbidity

Answer: B

Which chemical is commonly added to raise pH and reduce corrosion?

A. Sulfuric acid

B. Lime, or $Ca(OH)_2$

C. Hydrochloric acid

D. Nitric acid

Answer: B

What pH range is considered optimal for drinking water to minimize corrosion?

A. 4.0–5.0

B. 5.5–6.0

C. 7.0–7.8, neutral to slightly alkaline

D. 9.0–10.0

Answer: C

Which chemical is used to add alkalinity when using alum coagulant?

A. Sulfuric acid

B. Soda ash, or Na_2CO_3

C. Hydrochloric acid

D. Nitric acid

Answer: B

What happens if pH drops below 5.5 during alum coagulation?

A. Better floc formation

B. Poor floc formation and corrosion risks

C. No effect

D. Increased chlorine residual

Answer: B

Which chemical acts as a base to raise pH?

A. Sulfuric acid, or H_2SO_4

B. Caustic soda, or NaOH

C. Hydrochloric acid, or HCl

D. Nitric acid, or HNO_3

Answer: B

When using caustic soda, what minimum alkalinity is needed for stable pH?

A. 5 mg/L

B. 10 mg/L

C. 20 mg/L

D. 50 mg/L

Answer: C

Which chemical removes non-carbonate hardness during precipitation softening?

A. Lime, or $Ca(OH)_2$

B. Soda ash, or Na_2CO_3

C. Alum

D. Chlorine

Answer: B

What protective film do orthophosphates form on pipe walls?

A. Chlorine film

B. Phosphate film that inhibits metal leaching

C. Carbon film

D. Sulfate film

Answer: B

What is the minimum hardness that lime-soda ash softening can achieve?

A. 10–20 mg/L

B. 30–40 mg/L

C. 50–60 mg/L

D. 100 mg/L

Answer: B

Which condition causes increased formation of THMs, or trihalomethanes?

A. Low hardness

B. Low temperature

C. Low pH

D. High pH

Answer: D

Which statement about caustic soda vs. soda ash is true?

A. Both are acids

B. Caustic soda is a weaker base than soda ash

C. Caustic soda is a stronger base than soda ash

D. Both are neutral

Answer: C

What chemical is used to restore carbonate balance in softened water?

A. Lime

B. Carbon dioxide, or CO_2

C. Soda ash

D. Alum

Answer: B

Which condition increases corrosion risk in water systems?

A. High pH, above 8.5

B. Low pH, below 6.5

C. High alkalinity

D. High temperature

Answer: B

What is the #1 cause of taste and odor complaints in water systems?

A. High chlorine

B. Cross connections

C. Algae growth

D. Corroded pipes

Answer: B

Which oxidant is least effective for taste and odor control?

A. Chlorine

B. Potassium permanganate

C. Chlorine dioxide

D. Ozone

Answer: B

What is the maximum acceptable Threshold Odor Number, or TON, for drinking water?

A. 1

B. 2

C. 3

D. 5

Answer: C

Which method is preferred when taste and odor are moderate and infrequent?

A. Aeration

B. Powdered activated carbon, or PAC

C. Ozone

D. UV

Answer: B

What is the best point of application for powdered activated carbon, or PAC?

A. After filtration

B. In filters after coagulation or at the raw water intake

C. After chlorine only

D. At the customer endpoint

Answer: B

Why is chlorine dioxide better than regular chlorine for algae odors?

A. It is more expensive

B. It does not form chlorinous taste or THMs

C. It is less effective

D. It requires higher pH

Answer: B

Which chemical is used as an algaecide for surface water?

A. Alum

B. Copper sulfate

C. Soda ash

D. Lime

Answer: B

What odor is associated with decaying organic matter?

A. Metallic

B. Muddy or earthy

C. Chlorinous

D. Rotten egg

Answer: B

Which two chemicals are strong bases that raise pH?

A. Sulfuric acid and hydrochloric acid

B. Lime and caustic soda

C. Nitric acid and sulfuric acid

D. Alum and soda ash

Answer: B

Which condition is most common for algae-related taste and odor problems?

A. Winter, cold water

B. Summer/fall, warm water

C. Any season equally

D. Only during rain

Answer: B

CHAPTER 8

LABORATORY PROCEDURES AND MONITORING

Role of Laboratory Work in exam Operations

At the T2 level, operators must understand how lab data supports regulatory compliance, plant optimization, and troubleshooting.

Lab tests verify compliance with primary standards, such as turbidity and disinfectant residual, and secondary standards, such as hardness or taste.

Routine monitoring, including daily or continuous tests like turbidity and chlorine residual, is used to adjust processes such as coagulation, filtration, and disinfection.

Operators must know which tests are required at each treatment step, including raw, settled, filtered, and finished water, and understand the purpose of each test.

SAMPLE TYPES AND LOCATIONS

Grab vs. Composite Samples

A grab sample is a single sample taken at a specific time and location. It is used for parameters that change quickly, such as pH, chlorine residual, turbidity, and coliforms.

A composite sample is a combination of multiple samples collected over a period of time or at multiple points. Composite samples are more common in wastewater or for slowly changing parameters, such as BOD or some chemical constituents.

Plant Locations

Plant sampling locations include:

- Raw water: used to evaluate source quality and dosage needs.
- Settled water: used to evaluate coagulation, flocculation, and sedimentation performance.
- Filtered water: used to evaluate filter performance and turbidity removal.
- Finished water: used to confirm that water leaving the plant meets standards.

DISTRIBUTION LOCATIONS

Distribution samples may be collected from system taps, reservoirs, or dead ends. These samples are used to monitor chlorine residual, coliforms, and water quality changes in the system.

Sample Collection and Preservation

General Practices

Operators should flush the sample tap to remove stagnant water and remove aerators, strainers, or hoses when required.

Use the correct clean container for each sample. Sterile bottles are used for bacteriological samples, while clean plastic or glass containers may be used for routine chemistry.

Each bottle should be labeled properly with the date, time, location, sample type, and sampler initials.

PRESERVATION AND TIMING

Field tests such as pH, turbidity, and chlorine residual should be measured immediately or as soon as possible because these values can change quickly.

Some lab tests, such as metals or nutrients, require preservatives, refrigeration, and specific holding times set by the lab or approved methods.

Bacteriological samples must be cooled and delivered to a certified lab within the required time limits. Improper holding time can invalidate the result.

PH MEASUREMENT

pH affects corrosion, disinfection effectiveness, and chemical dosing. For T2, operators must know both the procedure and the operational meaning of pH.

Equipment and Calibration

Use a pH meter with a glass electrode and appropriate buffer solutions, commonly pH 4, 7, and 10.

Calibrate the meter at least daily, or according to the SOP, using two buffers that bracket the expected sample pH.

Basic Procedure

Rinse the electrode with distilled water between buffers and between samples.

Immerse the electrode in the sample, allow the reading to stabilize, and record the pH and temperature.

Measure pH promptly after collection because pH can drift over time.

Operational Significance

Lower pH can increase corrosivity and affect disinfection efficiency.

Higher pH can decrease the effectiveness of chlorine disinfection but may reduce corrosion.

ALKALINITY AND HARDNESS

Alkalinity and hardness are reported as mg/L as $CaCO_3$ and are related to buffering capacity and scaling potential.

Alkalinity

Alkalinity is the capacity of water to neutralize acid. It mainly comes from bicarbonates, carbonates, and hydroxides.

It is measured by titrating a known volume of sample with a standard acid to specific pH endpoints, using indicators or a pH meter.

Hardness

Hardness is primarily caused by calcium and magnesium. It affects scaling and soap consumption.

Hardness is usually measured by EDTA titration with an indicator, such as Eriochrome Black T.

TURBIDITY

Turbidity is a key regulatory parameter and an important process control indicator.

Measurement

Turbidity is measured in NTU, or Nephelometric Turbidity Units, using a calibrated turbidimeter. Samples should be gently mixed without air bubbles and analyzed as soon as possible.

Regulatory and Process Significance

Low turbidity in finished water is essential for effective disinfection and regulatory compliance. Turbidity in settled and filtered water is used to evaluate coagulation, flocculation, sedimentation, and filter performance.

Sudden increases in turbidity can indicate filter breakthrough, poor coagulation, or other process failures.

CHLORINE RESIDUAL

Chlorine residual is critical for disinfection and distribution system protection.

Types of Residual

Free chlorine residual is hypochlorous acid and hypochlorite ion available for disinfection. Combined chlorine residual is chloramines formed when chlorine reacts with ammonia. These are less powerful disinfectants but are more stable.

Total chlorine residual is the sum of free and combined residual.

MEASUREMENT, OR DPD METHOD

Collect a fresh grab sample in a clean container.

Add DPD reagent, observe color development, and compare it with a color comparator or measure it with a photometer.

Measure promptly because chlorine decays over time.

OPERATIONAL SIGNIFICANCE

Operators must maintain a minimum disinfectant residual at the plant effluent and throughout the distribution system to protect against microbial contamination. Excessive residual can cause taste and odor complaints.

COLIFORM SAMPLING OVERVIEW

Coliform testing is done by certified labs, but operators must know how to collect, handle, and interpret results at a basic level.

Sampling

Use sterile bottles containing a dechlorinating agent to neutralize chlorine.

Flush the tap, remove attachments, and follow SOPs for tap disinfection if required.

Avoid contamination by not touching the inside of caps or bottles and by not laying caps down.

Handling and Interpretation

Keep samples cooled and deliver them to the lab within the required holding time.

Understand the purpose of routine total coliform monitoring in the distribution system. Know that detection of coliforms can trigger repeat sampling and investigation requirements under the Total Coliform Rule.

CALIBRATION AND QUALITY CONTROL

Calibration

Instruments must be calibrated using certified standards at specified intervals.

Calibration records should be maintained, including the date, standards used, and results.

Blanks and Duplicates

Blanks, usually distilled or deionized water, help detect contamination from glassware, reagents, or the environment.

Duplicate samples are used to check precision. Large differences between duplicate results may suggest procedural problems or instrument issues.

Known standards or check samples can be analyzed to make sure tests are accurate. If control results are outside acceptable ranges, operators should investigate and correct the problem before trusting sample results.

Control Samples

RECORDKEEPING AND REPORTING FOR EXAM

Good records are required for regulatory compliance and plant management.

Lab Records

Each measurement should include:

- Sample ID.
- Date and time.
- Location.
- Test method.
- Result.
- Units.
- Operator initials.

Calibration logs and maintenance records for instruments should also be kept.

Regulatory Reporting

Data used for compliance reports, such as turbidity, disinfectant residual, and coliform results, must be complete and accurate.

Missing or incorrect data can be treated as violations.

Operational Use

Operators should trend data over time to detect gradual changes in source water, treatment performance, or distribution system conditions.

Corrective actions should be documented when results exceed internal or regulatory limits.

EXAM PREPARATION FOR CHAPTER 8

Which of the following is the best description of a grab sample?

A. A mixture of equal volumes collected every hour for 24 hours

B. A single sample collected at a specific time and place

C. A sample that is automatically collected by a composite sampler

D. A sample collected only from the clearwell

Answer: B

Which parameter is most commonly collected as a grab sample in a drinking water plant?

A. Weekly TDS

B. Monthly hardness

C. Chlorine residual in distribution

D. Annual lead and copper

Answer: C

In a surface water plant, which location is best for evaluating filter performance?

A. Raw water intake

B. Settled water

C. Individual filter effluent

D. Clearwell discharge only

Answer: C

A sample collected at multiple times over an 8-hour period and combined into one bottle is called:

A. Grab sample

B. Composite sample

C. Duplicate sample

D. Split sample

Answer: B

Which of the following best describes a distribution system sample?

A. Sample collected only at the plant clearwell

B. Sample taken from a finished water storage tank overflow

C. Sample taken from system taps or hydrants away from the plant

D. Sample taken from raw water to check source quality

Answer: C

Why is it important to flush a sampling tap before collecting a distribution grab sample?

A. To cool the tap so the operator can touch it

B. To ensure water from the main reaches the tap

C. To disinfect the tap threads

D. To increase chlorine demand

Answer: B

Which sample type is most appropriate for routine pH checks at a treatment plant?

A. Weekly composite sample

B. Monthly composite sample

C. Daily grab sample

D. Quarterly integrated sample

Answer: C

Which action is most important when preserving a bacteriological, or coliform, sample?

A. Fill bottle completely and freeze immediately

B. Use a bottle containing acid to lower pH

C. Use a sterile bottle with dechlorinating agent and keep it cooled

D. Add chlorine to the bottle before sampling

Answer: C

A plant operator collects a metals sample for a contract lab. What simple preservation step is often required in the field?

A. Add sodium hydroxide and store warm

B. Add acid, per instructions, and cool to about 4°C

C. Add dechlorinating agent and heat sample

D. Add chlorine to maintain residual

Answer: B

Which of the following problems will most likely make a coliform sample invalid?

A. Sample collected as a grab

B. Delivered to the lab the same day on ice

C. Collected in a non-sterile bottle without dechlorinating agent

D. Taken after flushing the tap

Answer: C

Why should pH be measured as soon as possible after sample collection?

A. pH never changes with time

B. pH may change as gases enter or leave the water

C. pH meters only work in the field

D. pH buffers are not available in the lab

Answer: B

A pH meter is calibrated with pH 7 and pH 10 buffers before use. This is an example of:

A. Duplicate sampling

B. Quality assurance

C. Chain of custody

D. Composite sampling

Answer: B

13. Which statement best describes alkalinity?

A. Measure of total dissolved solids

B. Capacity of water to neutralize acid

C. Measure of the total hardness

D. Indicator of dissolved oxygen

Answer: B

In most drinking water lab procedures, alkalinity is reported as:

A. mg/L as NaCl

B. mg/L as $CaCO_3$

C. mg/L as Cl_2

D. mg/L as Fe

Answer: B

Which ions are primarily responsible for water hardness in drinking water?

A. Sodium and chloride

B. Calcium and magnesium

C. Iron and manganese

D. Nitrate and sulfate

Answer: B

A hardness titration is being performed. The indicator color does not change even after adding a large amount of titrant. What is the most appropriate operator action?

A. Ignore the result and record zero hardness

B. Check reagents, repeat the test, and review the procedure

C. Dilute the sample until the color changes

D. Add more indicator until the color changes

Answer: B

High hardness in treated water is most likely to cause:

A. Pipe corrosion and low pH

B. Taste and odor problems only

C. Scaling in pipes and equipment

D. Low turbidity readings

Answer: C

Which of the following best explains why alkalinity is important for corrosion control?

A. High alkalinity always increases corrosion

B. Alkalinity buffers pH changes and helps stabilize water

C. Alkalinity only affects color

D. Alkalinity only affects taste and odor

Answer: B

Turbidity in finished water is primarily used as an indicator of:

A. Taste and odor

B. Bacterial growth only

C. Suspended particles and potential pathogens

D. Hardness and alkalinity

Answer: C

Which unit is used to report turbidity?

A. mg/L

B. NTU

C. ppm as $CaCO_3$

D. mg/L as Cl_2

Answer: B

sudden increase in filtered water turbidity usually indicates:

A. Improved coagulation

B. Filter breakthrough or process problem

C. Low hardness

D. Lower chlorine demand

Answer: B

Before measuring turbidity, the operator should:

A. Shake the sample vigorously to aerate it

B. Wipe the sample vial with a lint-free cloth and remove bubbles

C. Add chlorine to clarify the sample

D. Add EDTA to soften the water

Answer: B

Why should turbidity standards be used regularly with a turbidimeter?

A. To clean the glass vials

B. To check calibration and instrument accuracy

C. To reduce turbidity in samples

D. To increase sample pH Answer: B

A plant's filtered water turbidity is trending upward over several hours. What is the best first step for the operator?

A. Ignore the trend until it exceeds limits

B. Check filters, coagulant dose, and backwash schedule

C. Immediately shut down the plant

D. Increase distribution chlorine dose

Answer: B

Which method is commonly used by operators to measure chlorine residual?

A. BOD test

B. Jar test

C. DPD colorimetric method

D. DO probe

Answer: C

Free chlorine residual is best described as:

A. Chlorine combined with ammonia only

B. Chlorine combined with organic compounds

C. Chlorine available for disinfection, including hypochlorous acid and hypochlorite

D. Chlorine that has no disinfecting ability

Answer: C

Why must chlorine residual be measured quickly after collecting a sample?

A. Residual always increases with time

B. Chlorine can decay and give a falsely low result

C. Chlorine will change to turbidity

D. pH will always increase

Answer: B

A distribution sample shows no detectable chlorine residual. What is the best immediate operator response?

A. Assume meter failure and ignore the result

B. Recheck the sample and instrument, then investigate the cause if confirmed

C. Increase fluoride dose

D. Lower pump speed

Answer: B

High chlorine residual in distribution is most likely to cause:

A. Taste and odor complaints

B. Coliform contamination

C. High turbidity

D. Higher hardness

Answer: A

Which relationship is most important when considering disinfection effectiveness?

A. pH and hardness only

B. CT = disinfectant concentration × contact time

C. Turbidity and hardness

D. Flow and headloss

Answer: B

The main reason for routine total coliform sampling in a distribution system is to:

A. Measure TDS

B. Check corrosion control

C. Monitor for potential fecal contamination

D. Measure hardness

Answer: C

Which of the following best describes the operator's role when a total coliform-positive sample is reported?

A. Ignore the result unless customers complain

B. Collect required repeat and check samples and assist in investigation

C. Increase hardness

D. Decrease pH below 6.0

Answer: B

Which action will help avoid contamination when collecting a coliform sample?

A. Leaving the aerator on the tap

B. Touching the inside of the bottle cap

C. Holding the bottle under the tap without touching the rim

D. Filling the bottle halfway only

Answer: C

A coliform sample is taken from a hose bib in a warehouse. Which is the biggest concern with this sampling location?

A. Water pressure is too high

B. Hose and bib may not represent mainline water quality

C. Tap cannot be flushed

D. D. Tap cannot be labeled

Answer: B

A sample is collected in the correct sterile bottle and cooled but arrives at the lab after the allowed holding time. How should this sample be treated?

A. As valid and used for compliance

B. As invalid; a new sample may be required

C. As a composite sample

D. As a hardness test only

Answer: B

The main purpose of calibration is to:

A. Make data look better

B. Adjust an instrument so readings match known standards

C. Reduce turbidity in samples

D. Eliminate the need for standards

Answer: B

A blank sample is best described as:

A. A sample with a very high concentration

B. A sample collected from the clearwell

C. A sample of distilled or deionized water used to check contamination

D. A sample with unknown origin

Answer: C

Duplicate samples are collected primarily to:

A. Measure different parameters

B. Check precision and repeatability of results

C. Increase chlorine demand

D. Reduce lab workload

Answer: B

Which of the following information should always be recorded on a lab bench sheet or log for each sample?

A. Operator's home address

B. Date, time, location, test performed, and result

C. Only the test result

D. Only the plant name

Answer: B

Why is good recordkeeping important for T2 operators?

A. Records are only for operator practice

B. Records are required to document compliance and support operational decisions

C. Records replace the need for calibration

D. Records are only needed during emergencies

Answer: B

CHAPTER 9

PROCESS CONTROL, INSTRUMENTATION, SCADA, SAFETY, AND WATER REGULATION

Process Control Key Parameters in Water Treatment

The key process control parameters in water treatment include pH, turbidity, disinfectant residuals, flows, and levels. These measurements help operators keep the plant stable, safe, and compliant.

PH

pH matters because it controls coagulation effectiveness, corrosion, scaling, and disinfection efficiency.

Operator actions may include adjusting acid or caustic feed, changing lime or soda ash dose, or adjusting corrosion-control chemicals based on pH trends and lab checks.

TURBIDITY

Turbidity is a direct indicator of suspended solids and filter or clarifier performance. High turbidity can reduce disinfection effectiveness.

Operator actions may include changing coagulant or polymer dose, mixing intensity, filter backwash frequency, and filter-to-waste time based on turbidity readings and trends.

DISINFECTANT RESIDUALS

Disinfectant residuals help ensure adequate pathogen inactivation, CT, and a protective residual in the distribution system.

Typical control points include:

- Post-disinfection residual leaving the clearwell or plant.
- Residual at distribution system entry points and remote locations to check decay and water age. Operator actions may include adjusting disinfectant feed rate, contact time by changing flow or tank level, and sometimes pH or ammonia for chloramination to maintain the target residual profile.

FLOWS

Flows matter because they affect detention times, chemical dosage rates, and filter loading rates.

Operator actions include:

- Adjusting pump speeds, pump starts, valve positions, and chemical feed settings.
- Feeding chemicals in proportion to flow.
- Maintaining required contact times and filter loading limits.

LEVELS

Levels in tanks, clearwells, and reservoirs matter because they protect pump suction head, maintain required contact volume, and balance supply with demand.

Typical control points include:

- Clearwell level for CT, contact time, and stable disinfection.
- Filtered water or finished water storage tanks, standpipes, and reservoirs for system pressure and emergency storage.

Operator actions include using level setpoints to start and stop pumps, modulate flows, and maintain required minimum and maximum levels.

BASIC INSTRUMENTATION

Basic instrumentation means using sensors and controllers, often through SCADA, to measure flow, level, pressure, and other process conditions. These systems help automatically adjust pumps and valves to keep the process stable and safe.

Instrumentation means the devices that measure process variables and send signals to displays, controllers, or SCADA.

Control means using those signals to adjust pumps, valves, and other equipment to maintain desired setpoints.

Key Ideas

- Measured or controlled variable: the value operators care about, such as flow, level, or pressure. • Sensor or transmitter: converts the physical variable into a standard signal, such as 4–20 mA or a digital signal.
- Controller: compares the measurement to the setpoint and decides how to move a valve or pump.

For T2, operators mainly need to understand what common instruments do and how they relate to SCADA and operations.

FLOW METERS

Flow meters measure how much water is moving in a pipe or channel, often in gpm or MGD. Why They Matter

Flow meters help operators:

- Track source, plant, and distribution flows for operations and reporting.
- Calculate daily production.
- Control processes such as filter loading rates and chemical dosing based on flow.

Common Types

Common flow meters include:

- Turbine or propeller meters: use a mechanical rotor in the pipe. Rotor speed is proportional to flow.
- Magnetic, or mag, meters: use conductivity and an electromagnetic field. They have no moving parts and are common in water plants.
- Venturi or differential-pressure meters: measure pressure drop across a constriction to estimate flow.
- Open-channel devices: use flumes or weirs with level sensors to convert head, or depth, into flow.

OPERATOR POINTS

Flow signals often go to SCADA as continuous 4–20 mA or digital signals.

Flow totalizers accumulate volume over time and are used for production reports and chemical dosage calculations.

LEVEL SENSORS

Level instruments measure how full tanks, clearwells, or reservoirs are.

Why They Matter

Level sensors help:

- Protect pumps from running on empty.
- Prevent overflows.
- Maintain stable system pressures.
- Provide level data to SCADA for automatic pump start/stop and alarm functions.

COMMON LEVEL MEASUREMENT METHODS

Common level measurement methods include:

- Float switches or mechanical floats: simple on/off level detection, often used for pump control.
- Pressure or hydrostatic level sensors: measure pressure and convert it to water level.
- Ultrasonic or radar level sensors: send a signal from the top of the tank and measure the echo time to estimate level.
- Bubbler systems: measure back pressure in a tube with air bubbling into the liquid.

OPERATOR POINTS

Level setpoints, such as "start pump at 10 ft" and "stop pump at 14 ft," are used by controllers or PLCs to control pumps automatically.

Faulty level sensors can cause pumps to short-cycle, fail to start, or cause overflows, so alarms are important.

BASIC CONTROL CONCEPTS

Most water-plant control is closed-loop feedback control.

Core Ideas

- Setpoint: the desired value, such as clearwell level at 14 ft or plant flow at 5 MGD.
- Process variable, or PV: the actual measured value from the instrument.
- Controller output: the signal sent to adjust pump speed, valve position, or on/off state to bring the process variable toward the setpoint.

Example

If the clearwell level drops below the low-level setpoint, the controller starts the high-service pump. The level rises toward the setpoint, and the pump stops when the high-level setpoint is reached.

SCADA OVERVIEW

SCADA means Supervisory Control and Data Acquisition. It is the combination of hardware and software that lets operators monitor and control the system from a central location.

SCADA helps operators watch alarms, trends, flows, levels, pressures, pump status, valve status, and other plant information.

BASIC LAB WORK

Basic lab work includes the core operator tests used in T2: pH, alkalinity, hardness, turbidity, chlorine residual, and coliform sampling. These tests show whether treatment is working and whether the water is safe.

PH

pH measures how acidic or basic water is. It affects corrosion, disinfection, and coagulation. Drinking water is usually kept between about 6.5 and 8.5 for corrosion control and customer acceptance.

Key points for T2:

- pH is a measure of hydrogen ion concentration.
- Low pH means acidic water.
- High pH means basic water.
- Low pH can make water more corrosive and dissolve metals from pipes.
- High pH can reduce chlorine disinfection efficiency and may give water a slippery feel. Operator-level test methods include using a pH meter with proper buffer calibration or a color comparator.

pH should be measured as soon as possible after sampling because CO_2 exchange can change the pH.

ALKALINITY

Alkalinity is the water's capacity to neutralize acids. It mainly comes from bicarbonate, carbonate, and hydroxide ions. It is usually reported as mg/L as $CaCO_3$ and is important for pH stability and corrosion control.

Key points:

- Alkalinity acts as a buffer and resists rapid pH changes.
- It helps maintain stable conditions for corrosion control and chemical treatment.
- Low alkalinity means pH can swing easily and corrosion risk may increase.
- Moderate alkalinity helps keep pH in the target range for distribution.

Alkalinity is usually measured by standard acid titration to specific pH endpoints using indicators or a pH meter.

HARDNESS

Hardness is primarily the concentration of calcium and magnesium in water, usually expressed as mg/L as $CaCO_3$. It affects scaling, customer complaints, soap use, and sometimes treatment choices.

Types of hardness include:

- Temporary hardness: associated with bicarbonates and can be reduced by boiling or lime softening.
- Permanent hardness: associated with sulfates and chlorides and is more difficult to remove. Hardness is usually measured by EDTA titration with indicators, such as Eriochrome Black T, to a color endpoint.

High hardness can cause scale in pipes, heaters, and equipment. Very low hardness may be linked with more corrosive water if alkalinity is also low.

TURBIDITY

Turbidity is a measure of the cloudiness of water caused by suspended particles and colloids. It is a key indicator of treatment performance and potential pathogen risk.

Key points:

- Turbidity is measured in NTU using a nephelometric turbidimeter.
- Operators commonly check turbidity on raw, settled, filtered, and finished water.
- High turbidity increases filter loading and can shield microorganisms from disinfectant. • Regulatory standards usually require very low filtered-water turbidity for surface water supplies.

Operator practices include calibrating or verifying the meter with standards and keeping vials clean and bubble-free.

Operators should also trend turbidity to catch filter breakthrough or poor coagulation early.

CHLORINE RESIDUAL

Chlorine residual is the amount of chlorine remaining in water after a given contact time. It is available for disinfection and distribution protection.

Types of chlorine residual include:

- Free chlorine residual: chlorine forms that directly disinfect, such as hypochlorous acid and hypochlorite.
- Combined residual: chloramines formed when chlorine reacts with ammonia.
- Total residual: the sum of free and combined forms.

The common operator test method is the DPD colorimetric method using a comparator or photometer.

A minimum residual is required in finished water and throughout the distribution system to protect against microbial contamination. Excessive residual can cause taste and odor complaints, but too low a residual may indicate regrowth or contamination.

COLIFORM SAMPLING OVERVIEW

Coliforms, especially total coliforms and E. coli, are indicator bacteria used to assess microbial safety of drinking water. Proper sampling technique is essential to avoid false positives and to make sure results are valid.

Purpose

Total coliforms indicate possible contamination from sewage, animals, or environmental sources. Routine monitoring in distribution is required by drinking water rules to verify system integrity.

SAMPLING BASICS

Use sterile bottles provided by a certified lab, usually containing a dechlorinating agent. Choose appropriate taps that are clean, in good repair, cold-water only, and without hoses or point-of-use devices.

Flush the tap to bring mainline water, usually for several minutes until temperature and residual stabilize.

Avoid contaminating the sample by not touching the inside of the cap or bottle and by controlling flow to avoid splashing.

HANDLING

Keep samples cooled and deliver them to the lab within the required holding time, often within 24 to 30 hours.

Use accredited labs and follow chain-of-custody and documentation requirements.

SAMPLE COLLECTION AND PRESERVATION

Sample collection and preservation are about getting a representative sample from the right place and keeping it unchanged until testing.

WHY SAMPLING AND PRESERVATION MATTER?

Sampling errors can ruin even good lab work, so operators must collect and preserve samples correctly.

The sample must represent the water being checked, whether it is from a process step or distribution location.

Preservation and handling must prevent changes such as chlorine loss, bacterial growth, or chemical reactions before testing.

GRAB SAMPLES

A grab sample is a single sample collected at one point in time and space. It is a snapshot of conditions at that moment.

Operator-level points:

- Used for parameters that can change quickly, such as pH, chlorine residual, turbidity, and coliforms.
- Usually collected by hand directly into the sample container or with a sampling device.
- Results apply only to the time and place of collection, not an average over time.

Examples in drinking water include:

- Chlorine residual at a distribution tap.
- Filter effluent turbidity at a specific time.
- Coliform sample from a designated site.

COMPOSITE SAMPLES

A composite sample combines multiple smaller samples, called aliquots, collected over time or at different points into one mixed sample. It represents average conditions.

Operator-level points:

- Common in wastewater or where regulations call for a time-weighted or flow-weighted average.
- Collected manually or automatically with a composite sampler.
- Good for average loadings or constituents that vary slowly.
- Not used for highly time-sensitive tests.

Composite samples are less common for basic drinking water operator lab tests. Drinking water tests for pH, turbidity, chlorine residual, and coliform are usually grab samples because conditions can change quickly and rules often require instant checks.

CHOOSING GRAB VS. COMPOSITE

Use grab samples when:

- You need conditions at a specific moment.
- The parameter changes rapidly or is time dependent.
- The regulation specifically requires a grab sample. Use composite samples when:
- You need an average over a time period.
- You are doing mass loading or average concentration calculations.

PLANT VS. DISTRIBUTION SAMPLING

Plant, or Process, Sampling

Plant samples are taken at treatment steps to monitor and control processes.

Common locations include:

- Raw water: source quality, including turbidity, pH, alkalinity, and temperature.
- Settled water: coagulation, flocculation, and sedimentation performance.
- Filtered water: filter performance and pathogen barrier verification.
- Finished water: water leaving the plant, checked for turbidity, residual, and pH. Plant samples are used mainly for process control, such as adjusting coagulant dose, filter backwashing, and chlorine dosing.

They are often collected as grab samples at sampling taps on each process unit.

DISTRIBUTION SAMPLING

Distribution samples are taken out in the system, such as taps, hydrants, and reservoirs. They show what customers receive and help check for contamination.

Common uses include:

- Chlorine residual checks at different points in the system.
- Routine coliform monitoring at designated sites.
- Checking secondary issues such as taste, odor, and color at customer locations.

Sample taps should be clean, free of hoses or aerators, and representative of mainline water. Flushing is critical to bring fresh water from the main and avoid stagnant plumbing water.

GENERAL SAMPLING TECHNIQUE

No matter where or what operators sample, some basic practices apply.

Basic Steps

Select a representative location based on the monitoring plan, whether plant or distribution.

Flush the sample tap long enough to stabilize temperature and residual, often several minutes.

Use the correct container:

- Sterile, dechlorinated bottles for bacteriological samples.
- Clean plastic or glass bottles for routine chemistry. Avoid contamination:
- Do not touch the inside of the cap or bottle.
- Do not lay caps down.
- Do not allow backsplash into the bottle.

Label the sample immediately with date, time, location, and sampler initials.

PRESERVATION BASICS AND HOLDING TIMES

Preservation means stopping or slowing changes in the sample between collection and analysis.

General Principles

Time matters. Analyze samples as soon as possible. Some tests must be run immediately, including pH, chlorine residual, and turbidity.

Temperature matters. Many samples are preserved by cooling to about 4°C on ice in a cooler to slow biological and chemical reactions.

Chemicals may be needed. Some parameters require chemical preservatives such as acids, bases, or dechlorinating agents added to bottles before or at sampling.

- pH, chlorine residual, and turbidity should be tested immediately or very soon.
- Coliform samples require sterile, dechlorinated bottles and should be kept cold and delivered within the allowed holding time.
- Metals and some chemical parameters may need acidification, cooling, and method-specific holding times.

DOCUMENTATION AND CHAIN OF CUSTODY

Good documentation proves that samples were handled correctly.

Operator Responsibilities

Operators should fill out field forms or chain-of-custody forms with:

- Location.
- Date and time.
- Sampler name or initials.
- Preservatives used.
- Field measurements taken.

Sample identification should stay consistent from field to lab report so results can be tied to specific locations and times.

QA/QC at the Operator Level

QA/QC shows that data is trustworthy. It shows that instruments read correctly, samples are not contaminated, and results can be repeated.

Why QA/QC Matters

QA, or quality assurance, is the system of procedures that helps ensure results are reliable. QC, or quality control, is the checks operators actually run, such as calibrations, blanks, duplicates, and standards.

QA/QC matters because it:

- Protects operators from making bad decisions based on bad data.
- Supports regulatory compliance.
- Helps catch problems with instruments, reagents, or sampling before they cause violations.

CALIBRATION

Calibration means adjusting an instrument so its readings match known standards.

Typical Instruments

Typical instruments that need calibration or verification include:

- pH meters, calibrated with buffer solutions such as pH 4, 7, and 10.
- Turbidimeters, calibrated or verified with turbidity standards.
- Colorimeters or photometers, checked with prepared standards or built-in calibration kits.
- Flow meters, chlorine analyzers, and online sensors, often factory-calibrated but verified in the field.

OPERATOR RESPONSIBILITIES

Operators should:

- Calibrate at the frequency required by SOP, often daily or before use for lab meters.
- Use fresh, unexpired standards stored properly.
- Record calibration information.

CALIBRATION RECORDS SHOULD INCLUDE:

- Date and time.
- Standard values, such as 7.00 or 10.00.
- Instrument readings.
- Any adjustments made.

EXAM POINTS

Calibration is important because results must match known standards and be trusted.

If calibration fails, check or replace standards, clean the sensor, try again, and take the instrument out of service if it still fails.

A pH meter usually needs daily calibration, and turbidimeters or colorimeters may need verification depending on the SOP.

<u>Practical Examples</u>

If a pH meter reads 7.3 in a 7.0 buffer, the meter should be adjusted so it reads 7.0. If a turbidimeter verification standard of 0.3 NTU reads 0.8 NTU, operators should not trust production data until the instrument is corrected.

<u>Blanks</u>

Blanks are samples that should contain no analyte, often distilled or deionized water. They are used to check for contamination from reagents, glassware, or the environment.

<u>Common Blanks</u>

Common blanks at the operator level include:

- Reagent blanks for colorimetric tests, such as chlorine or iron.
- Distilled water blanks run through a titration or measurement procedure.
- Rinse blanks in some labs, though these are less common for basic operator labs.

<u>When Operators Use Blanks</u>

Blanks are used to confirm that reagents and glassware are clean.

They are also used to detect contamination. If the blank shows a significant reading, such as chlorine, color, or turbidity, there may be a contamination or reagent problem.

Operator Actions When a Blank Fails

If a blank fails, operators should:

- Stop using the affected result set.
- Check glassware cleanliness.
- Check reagent age and condition.
- Repeat the test with fresh reagents and clean containers.
- Document the issue and corrective actions.

EXAM ANGLES

The purpose of a blank is to detect contamination from reagents or equipment.

If the blank gives a high reading, results are suspect.

The best action is usually to identify and correct the source, then rerun the test.

Duplicates

Duplicates are two samples taken from the same location at nearly the same time, or one sample split into two and analyzed separately.

Purpose

Duplicates are used to check precision, or how close repeated measurements are to each other.

They also show whether sampling and measurement technique is consistent.

Types

Field duplicates are two separate grab samples collected one after the other from the same tap and tested independently.

Lab duplicates are one sample split into two portions in the lab and run separately.

How to Interpret Duplicates

If duplicate results are close, within an acceptable difference defined by the SOP, the method is precise.

If duplicate results differ significantly, operators should investigate technique, instrument stability, calibration, or mixing. The test may need to be repeated with better technique.

Example

Two duplicate turbidity samples from the same filter effluent show:

- 0.09 NTU and 0.10 NTU = good precision.
- 0.09 NTU and 0.35 NTU = poor precision, so check the instrument and method.

CONTROL STANDARDS

Control standards are known concentration samples, such as a chlorine standard or hardness standard. They are run like unknown samples to see if the test method gives the correct result. Purpose

Control standards verify accuracy, meaning how close results are to the true value.

They confirm that after calibration and good technique, the test still matches the expected value. Example

If a 1.00 NTU turbidity standard reads 0.98 NTU, it is within an acceptable range. If it reads 1.5 NTU, the instrument or procedure needs attention.

SIMPLE QA/QC ROUTINE

A realistic daily QA/QC routine for a T2-level operator may include checks before, during, and after testing.

BEFORE TESTING

- Calibrate the pH meter with two buffers.
- Verify the turbidity meter with low- and mid-range standards.
- Check chlorine reagents for expiration.
- Run a reagent blank when appropriate.

DURING TESTING

For key tests such as turbidity and chlorine, run:

- A blank if appropriate.
- An occasional duplicate sample from important locations.

AFTER TESTING

- Review differences between duplicates.
- Confirm standards or verification checks are within limits.
- Record all QA/QC in a log, including calibrations, blanks, duplicates, and standards.

SAFETY AND REGULATIONS

Safety and regulations for T2 focus on two main things: keeping operators safe on the job and making sure produced water meets federal and California requirements.

OPERATOR SAFETY

CHEMICAL HANDLING

Operators routinely handle hazardous chemicals such as chlorine, acids, and caustics. These chemicals can burn the skin, eyes, and lungs.

General rules include:

- Know the SDS, or Safety Data Sheet, for every chemical.
- Understand hazards, first aid, PPE, storage, and mixing rules.
- Use proper PPE, including chemical-resistant gloves, eye and face protection, protective footwear, and protective clothing.
- Always add acid to water, never water to acid, to reduce the risk of violent reactions and splashes.

CHLORINE, GAS AND LIQUID

Chlorine is a strong oxidizer and respiratory irritant. High concentrations can be fatal.

Key points:

- Chlorine gas is heavier than air and can collect in low areas.
- Use ventilation and fixed gas detection where required.
- Even small leaks can be dangerous.
- Leaks should be located only by safe methods described in SOPs, never with open flames.
- PPE may include chemical-resistant gloves, goggles, face shield, and appropriate respiratory protection where required.
- Never mix chlorine with acids or ammonia because this can release toxic gases.

ACIDS

Acids, such as hydrochloric or muriatic acid, can cause severe burns and release harmful fumes.

Key points:

- Wear gloves, splash goggles, face shield, and protective clothing.
- Avoid breathing fumes.
- Store acids separately from caustics and chlorine.
- Keep containers labeled and closed.
- Use secondary containment and proper ventilation in chemical rooms.

CAUSTICS

Caustics, such as sodium hydroxide, can cause deep tissue burns and eye damage.

Key points:

- Wear chemical-resistant gloves, goggles or face shield, and protective clothing.
- Caustics are slippery, so spills are both burn hazards and slip hazards.
- Neutralize and clean spills according to SOP.
- Store caustics away from acids and aluminum because hydrogen gas can form.

CONFINED SPACE SAFETY AND PPE

Many plant structures, such as tanks, wet wells, and vaults, may be confined spaces with hazardous atmospheres or entry and exit risks.

CONFINED SPACE BASICS

A permit-required confined space may have hazards such as:

- Low oxygen.
- Toxic gas.
- Flammable gas.
- Engulfment risk.
- Restricted entry or exit. Entry usually requires:
- Documented hazard assessment.
- Entry permit.
- Attendant.
- Rescue plan.
- Specific training under OSHA or Cal/OSHA rules.

Key Equipment

Key confined space equipment may include:

- Atmospheric testing with a 4-gas meter before and during entry.
- Ventilation, such as an air blower, where needed.
- Fall protection and retrieval systems, such as tripod, harness, and winch.
- Appropriate PPE, including helmet, eye protection, gloves, protective clothing, and respiratory
- protection where required.

EXAM-LEVEL UNDERSTANDING

Confined spaces require special procedures and training. Operators should not enter just because "the job needs to be done."

PPE BASICS

Personal Protective Equipment, or PPE, is the last line of defense. Engineering and administrative controls come first, but PPE is still essential.

COMMON PPE FOR WATER OPERATORS

Common PPE includes:

- Eye and face protection: safety glasses with side shields, chemical splash goggles, and face shields.
- Gloves: chemical-resistant gloves appropriate for acids, caustics, or chlorine.
- Protective clothing: coveralls, chemical aprons, boots, and high-visibility clothing as needed. • Respiratory protection: used only under a written program with fit testing when engineering controls are not enough and when required by risk assessment or regulation.

SAFE DRINKING WATER ACT

The U.S. Safe Drinking Water Act, or SDWA, is the main federal law that protects drinking water quality.

<u>Core Ideas</u>

EPA sets National Primary Drinking Water Regulations, or NPDWRs, for contaminants that can affect health.

These regulations are usually expressed as Maximum Contaminant Levels, or MCLs, or when an MCL is not practical, treatment techniques, or TTs.

Water systems must monitor, report, and correct violations. Public notification is required for certain violations.

MCLS AND TREATMENT TECHNIQUES

<u>Maximum Contaminant Levels</u>

An MCL is the maximum allowed level of a contaminant in drinking water delivered to users.

Key points:

- EPA first sets a non-enforceable health goal, called an MCLG.
- EPA then sets an enforceable MCL as low as feasible, considering technology and cost.
- MCLs cover inorganic chemicals, organics, microbiological contaminants, and radionuclides.
- Systems must monitor at required frequencies and meet the applicable limits.

TREATMENT TECHNIQUES

When it is not practical to directly set or measure an MCL, EPA can require a treatment technique instead.

Examples include:

- Filtration and disinfection requirements for surface water systems.
- Turbidity and log removal or inactivation requirements for pathogens.
- Optimized corrosion control under the Lead and Copper Rule.

A treatment technique is an enforceable procedure or performance level instead of a simple numeric contaminant limit.

LEAD AND COPPER RULE

The Lead and Copper Rule focuses on controlling lead and copper at the customer's tap through optimized corrosion control. It does not work like a simple MCL for tap samples.

Key Points

Systems collect tap samples at high-risk homes and compare results to action levels for lead and copper.

If action levels are exceeded, systems must take additional steps, such as:

- Optimizing corrosion control treatment.
- Providing public education.
- Replacing lead service lines when required by the current rule.

The Lead and Copper Rule is treated as a treatment technique rule because it focuses on controlling corrosion and plumbing leaching, not just meeting one simple MCL.

For T2, remember that lead and copper are controlled mainly by corrosion control, including pH, alkalinity, and inhibitors. Compliance is based on tap sampling and action levels.

CHAPTER 9 REVIEW

The main purpose of process control in a water treatment plant is to:

A. Reduce operator staffing

B. Maintain stable water quality and system performance

C. Eliminate the need for lab testing

D. Increase energy use to full capacity

Answer: B

In a typical control loop, the value being measured, such as flow or level, is called the:

A. Setpoint

B. Output signal

C. Process variable

D. Feedback controller

Answer: C

The desired value that the operator wants to maintain, such as 5 MGD flow, is called the:

A. Process variable

B. Setpoint

C. Disturbance

D. Differential

Answer: B

A mag meter installed on the plant effluent line is primarily used to:

A. Measure chlorine dose

B. Measure instantaneous and totalized plant flow

C. Measure pH and temperature

D. Measure sludge depth in a clarifier

Answer: B

Which type of signal is most commonly used for analog instrumentation in water plants?

A. A. 0–10 V DC

B. 120 V AC

C. 4–20 mA DC

D. 480 V AC

Answer: C

In a clearwell level control system, the level transmitter sends a signal to a controller, which turns pumps on and off. This is an example of:

A. Open-loop control

B. Closed-loop feedback control

C. Manual control only

D. Batch process

Answer: B

A float switch in a wet well is typically used to:

A. Measure pH

B. Turn pumps on or off at specific levels

C. Measure chlorine residual

D. Control filter backwash rate

Answer: B

A hydrostatic level sensor measures:

A. Water temperature

B. Water density

C. Pressure at a fixed depth and converts it to level

D. Air pressure above the tank

Answer: C

A sudden unexplained drop in high-service pump discharge pressure could indicate:

A. Pipe break or major leak

B. Increase in water temperature only

C. Calibration error on the pH meter

D. Higher alkalinity in raw water

Answer: A

Which instrument is best suited to measure the turbidity of filter effluent?

A. pH meter

B. Turbidimeter

C. Conductivity meter

D. Flow meter

Answer: B

A valve that is positioned automatically to maintain a set flow rate is controlled by:

A. SCADA historian

B. A flow control loop

C. An LCR meter

D. The SDWA

Answer: B

A plant operator notices that the pH meter reading drifts slowly upward during the day, even on a stable buffer. The best first action is to:

A. Ignore it and use the values

B. Recalibrate the pH meter and inspect the electrode

C. Increase chlorine dose

D. Shut down the plant

Answer: B

A differential pressure gauge across a filter is used to monitor:

A. Chlorine demand

B. Filter headloss and need for backwash

C. pH change across the filter

D. Temperature differences

Answer: B

In an open channel, a Parshall flume measures flow by relating:

A. Pressure to conductivity

B. Velocity to temperature

C. Water depth, or head, to a known flow equation

D. Chlorine residual to pH

Answer: C

Which of the following is a common sign that a flow meter may need calibration or maintenance?

A. Lab pH is stable

B. Flow reading is constant while pump speed and system demand change

C. Turbidity decreases

D. Distribution pressure is normal

Answer: B

A VFD, or variable frequency drive, on a pump is primarily used to:

A. Start generators

B. Adjust pump speed and control flow or pressure

C. Measure chlorine residual

D. Monitor turbidity

Answer: B

In a filter control system, the process variable might be:

A. Operator's shift schedule

B. Influent pH

C. Filter effluent turbidity

D. Color of the pump

Answer: C

A pressure transmitter on a distribution main provides information related directly to:

A. System pressure, or psi

B. Turbidity, or NTU

C. Chlorine residual, or mg/L

D. Alkalinity, or mg/L as $CaCO_3$

Answer: A

A sudden increase in plant flow with no change in setpoint could be caused by:

A. A leak in the distribution system

B. A pH meter calibration issue

C. Lower hardness in source water

D. Higher alkalinity

Answer: A

A temperature sensor installed on a clearwell helps operators because:

A. It controls all chemical dosing

B. Water temperature affects disinfection efficiency and distribution operations

C. Temperature does not matter in drinking water

D. It replaces lab testing

Answer: B

In a control loop, the final control element is typically:

A. Instrument air compressor

B. Sensor cable

C. Pump, valve, or damper being moved

D. SCADA historian

Answer: C

The most important reason for trending turbidity continuously on filter effluent is to:

A. Reduce chemical costs

B. Detect filter problems and protect public health

C. Measure pH changes

D. Monitor flow rate

Answer: B

A flow-paced chemical feed system adjusts chemical dose based on:

A. Operator vacation schedule

B. Raw water color only

C. Plant flow rate signal

D. Finished water temperature

Answer: C

When a level transmitter fails and shows a flat reading, the safest operator response is usually to:

A. Ignore the reading and continue automatic operation

B. Switch to manual control and verify level by another method

C. Increase pump speed

D. Lower pH

Answer: B

If a filter level sensor is reading much higher than actual while turbidity is still low, it is most likely:

A. A chlorine problem

B. A sensor calibration or malfunction issue

C. An MCL violation

D. A pH buffer issue

Answer: B

A pressure reducing valve, or PRV, in a distribution system is used primarily to:

A. Increase chlorine dose

B. Maintain downstream pressure within a safe range

C. Measure turbidity

D. Control pH in the main

Answer: B

A filter backwash is usually started when:

A. pH drops slightly

B. Flow rate increases

C. Headloss, or differential pressure, becomes too high or turbidity increases

D. Alkalinity increases

Answer: C

An online chlorine analyzer sending continuous data to SCADA allows operators to:

A. Eliminate all grab samples

B. Monitor residual trends and adjust dosing in near real time

C. Ignore regulatory sampling

D. Replace turbidity monitoring

Answer: B

If the SCADA flow totalizer shows 3.0 MGD but the billing records suggest much lower usage, one possible explanation is:

A. Negative alkalinity

B. Flow meter or SCADA scaling error

C. pH probe is dirty

D. Filters are oversized

Answer: B

Which of the following is the best definition of "setpoint" in process control?

A. The maximum allowable MCL value

B. The operator's desired operating value for a variable

C. The lowest possible reading of an instrument

D. The average of all daily samples

Answer: B

SCADA stands for:

A. Supervisory Control and Data Acquisition

B. System Control and Data Analysis

C. Standard Compliance and Data Access

D. Signal Control and Digital Automation

Answer: A

The main purpose of SCADA in a water system is to:

A. Replace all operators with computers

B. Monitor and control remote facilities and processes

C. Eliminate safety training

D. Replace lab analyses **Answer: B**

Which component of a SCADA system typically interfaces directly with field instruments and controls pumps and valves?

A. Historian

B. PLC or RTU

C. Modem

D. Printer

Answer: B

The part of SCADA that displays graphical screens, trends, and alarms to operators is called the:

A. HMI, or Human-Machine Interface

B. Historian

C. PLC

D. RTU

Answer: A

A SCADA historian is used to:

A. Run the chemical feed pumps

B. Store and retrieve historical data for trending and reports

C. Control chlorine gas rooms

D. Calibrate sensors

Answer: B

A communication failure alarm from a remote tank in SCADA means:

A. The tank is empty

B. The sensor is reading high

C. SCADA has lost contact with the remote site

D. There is always a chlorine leak

Answer: C

One advantage of SCADA for operators is:

A. It removes the need for on-site inspections

B. It provides real-time data and remote control for faster response

C. It automatically fixes mechanical problems

D. It replaces regulatory monitoring

Answer: B

A SCADA alarm for "high-high clearwell level" is usually configured to:

A. Turn off plant influent pumps to prevent overflow

B. Start additional chemical feed

C. Shut down the distribution system

D. Lower chlorine residual

Answer: A

When viewing SCADA trends, a gradually declining distribution pressure at night might indicate:

A. Normal diurnal pattern

B. Possible leak or pressure zone issue

C. Filter breakthrough

D. pH meter drift

Answer: B

Cybersecurity concerns for SCADA primarily involve:

A. Lab data integrity only

B. Unauthorized access or control of critical systems

C. Taste and odor complaints

D. Corrosion control

Answer: B

If SCADA incorrectly shows a pump running when it is actually off, the most likely cause is:

A. Operator math error

B. Incorrect status input wiring or failed status contact

C. High pH

D. Low residual

Answer: B

The best practice when SCADA and local panel indicators disagree about equipment status is to:

A. Trust SCADA only

B. Trust the local panel only

C. Verify physically in the field if safe and follow SOP

D. Ignore both

Answer: C

A remote start/stop command from SCADA:

A. Permanently overrides local safety interlocks

B. Should work within the limits of local controls and interlocks

C. Eliminates the need for local controls

D. Is not allowed in water plants

Answer: B

SCADA trends of chlorine residual and pH together can help operators:

A. Determine filter media type

B. Understand disinfection effectiveness and adjust dose

C. Calculate pump horsepower

D. Determine hardness

Answer: B

A SCADA alarm that is frequently acknowledged and reoccurs without investigation is:

A. Good practice

B. A sign of alarm fatigue and poor alarm management

C. Required by regulation

D. Needed to save energy

Answer: B

The best reference for hazards, PPE, and first aid for a specific chemical is the:

A. Operator's opinion

B. Safety Data Sheet, or SDS

C. SCADA screen

D. Lab notebook

Answer: B

When diluting acids, safe practice is to:

A. Add water to acid

B. Add acid to water

C. Mix acid and caustic together

D. Add acid to chlorine

Answer: B

Chlorine gas is:

A. Lighter than air and rises

B. Heavier than air and tends to settle in low areas

C. Non-toxic

D. Odorless

Answer: B

A small chlorine leak is suspected in a gas room. The operator does not have proper respiratory equipment. The correct action is to:

A. Enter the room quickly to check

B. Hold breath and search for the leak

C. Stay out, evacuate the area as required, and follow the emergency response plan

D. Open all doors and wait inside

Answer: C

Which PPE is most critical when handling corrosive acids or caustics?

A. Cloth gloves only

B. Chemical-resistant gloves, splash goggles, and face shield

C. Ear plugs only

D. Steel-toe boots only

Answer: B

Confined spaces may be hazardous because they:

A. Always have high turbidity

B. Can have low oxygen, toxic gases, and restricted entry/exit

C. Never contain chemicals

D. Are always safe if ventilated

Answer: B

Before entering a permit-required confined space, operators must:

A. Only check chlorine residual

B. Complete entry permit procedures, atmospheric testing, and follow the rescue plan

C. Ask a coworker for verbal permission

D. Rely on SCADA

Answer: B

A 4-gas meter is commonly used to monitor:

A. pH, alkalinity, hardness, and turbidity

B. Oxygen, flammable gas, and specific toxic gases

C. Hardness, alkalinity, and chlorine

D. MCLs for all contaminants

Answer: B

Lockout/tagout procedures are used to:

A. Increase plant flow

B. Protect workers from unexpected startup or release of energy

C. Calibrate instruments

D. Measure turbidity

Answer: B

A chemical storage room should have:

A. No ventilation

B. Proper ventilation, secondary containment, and clear labeling

C. Food storage

D. Unlabeled containers

Answer: B

Strong acids and strong caustics should be stored:

A. Together to save space

B. In the same open drain

C. Separately, to prevent dangerous reactions

D. In the same plastic tub

Answer: C

A splash of acid into the eye should be handled by:

A. Immediately flushing with water at an eyewash station for at least 15 minutes and seeking medical attention

B. Waiting for pain to pass

C. Applying chlorine solution

D. Using dry wipes only

Answer: A

Steel-toe boots in a treatment plant help protect against:

A. Chlorine gas exposure

B. Impact and crush hazards from equipment or containers

C. Noise

D. UV radiation

Answer: B

Wearing hearing protection is important in areas:

A. With very low light

B. With high noise levels from pumps and blowers

C. With high turbidity

D. With low chlorine residual

Answer: B

A fall protection harness and retrieval system are often required for:

A. Reading SCADA screens

B. Entering vertical tanks or deep vaults

C. Taking pH measurements

D. Collecting grab samples from faucets

Answer: B

The main reason to avoid mixing hypochlorite solution with acid is that it can:

A. Lower pH

B. Produce toxic chlorine gas

C. Increase hardness

D. Increase alkalinity

Answer: B

Good housekeeping in chemical storage and feed areas helps prevent:

A. SCADA malfunctions

B. Spills, accidents, and cross-contamination

C. Turbidity changes

D. pH drift

Answer: B

A chemical's NFPA diamond or GHS pictograms provide quick information about:

A. Plant capacity

B. Health, flammability, and reactivity hazards

C. MCLs

D. Turbidity standards

Answer: B

Respirators used in chlorine gas areas must be:

A. Shared without cleaning

B. Used only with dust filters

C. Selected, fit-tested, and maintained under a written respiratory protection program

D. Stored in the gas room

Answer: C

The first step when responding to any chemical exposure or spill should be to:

A. Ignore alarms

B. Ensure personal safety and follow the facility's emergency response plan

C. Pump more water

D. Shut down SCADA

Answer: B

A good example of administrative control for safety is:

A. A pump guard

B. A written SOP and training program

C. Chemical-resistant gloves

D. Ear plugs

Answer: B

A plant has a chlorine gas alarm system. If the alarm activates, the operator should:

A. Silence it and stay in the room

B. Evacuate per procedure and notify supervisors or emergency responders

C. Turn off the alarm and continue work

D. Add more chlorine

Answer: B

Safety showers and eyewash stations should be:

A. Hard to reach

B. Blocked by stored materials

C. Accessible, tested regularly, and clearly marked

D. Removed to save space

Answer: C

Safety signs and labels in a plant are used to:

A. Decorate walls

B. Communicate hazards, PPE requirements, and emergency information

C. Replace training

D. Meet only aesthetic standards

Answer: B

The most important safety attitude for operators is to:

A. Finish work quickly and ignore hazards

B. Assume all tasks are safe

C. Recognize hazards and follow procedures and PPE requirements consistently

D. Avoid reporting near misses

Answer: C

The main federal law governing public drinking water quality in the U.S. is the:

A. Clean Air Act

B. Safe Drinking Water Act

C. Clean Water Act

D. RCRA

Answer: B

National Primary Drinking Water Regulations are:

A. Non-enforceable guidelines

B. Legally enforceable standards expressed as MCLs or treatment techniques

C. Only state-level suggestions

D. Only for private wells

Answer: B

An MCL, or Maximum Contaminant Level, is:

A. A recommended goal

B. The maximum permissible level of a contaminant in water delivered to users

C. Always set at zero

D. Only used for secondary contaminants

Answer: B

An MCLG, or Maximum Contaminant Level Goal, is:

A. An enforceable limit

B. A non-enforceable health-based goal

C. Equal to the MCL in all cases

D. Only for secondary contaminants

Answer: B

For some contaminants, EPA uses a treatment technique instead of an MCL because:

A. MCLs are never allowed

B. It may be difficult to measure or regulate the contaminant directly

C. Operators prefer techniques

D. Treatment is optional

Answer: B

The Lead and Copper Rule, or LCR, is an example of a:

A. Secondary standard

B. Treatment technique rule focused on corrosion control and tap sampling

C. Turbidity rule

D. Disinfection byproduct rule

Answer: B

Under the LCR, compliance is primarily based on:

A. Plant effluent lead MCL

B. Tap samples at high-risk homes compared to action levels

C. Turbidity of finished water

D. Total chlorine residual only

Answer: B

An action level for lead or copper is:

A. The same as an MCL

B. A concentration that triggers treatment or other requirements if exceeded

C. A non-enforceable guideline

D. Only for small systems

Answer: B

Primary standards mostly address:

A. Aesthetic concerns like taste and odor

B. Health-related contaminants

C. Pipe materials

D. Painting requirements

Answer: B

Secondary standards are:

A. Enforceable in all states

B. Non-enforceable guidelines for aesthetic effects such as taste, odor, and color

C. Only used for microbiological contaminants

D. Not related to water quality

Answer: B

Under SDWA, states like California can:

A. Only use federal standards

B. Set more stringent drinking water standards than EPA

C. Ignore federal rules

D. Regulate only wastewater

Answer: B

Monitoring and reporting requirements are important because:

A. They are optional

B. They document compliance and allow detection of violations

C. They only affect lab staff

D. They replace operator certification

Answer: B

A violation of an MCL or treatment technique may require:

A. No action unless customers complain

B. Public notification and corrective action

C. SCADA upgrade only

D. Changing plant paint color

Answer: B

The Surface Water Treatment Rule requires surface water systems to:

A. Treat only for taste and odor

B. Filter and disinfect to achieve specified pathogen reductions

C. Use only groundwater

D. Eliminate turbidity monitoring

Answer: B

In California, drinking water operator certification and oversight are administered by:

A. EPA directly only

B. State Water Resources Control Board / Division of Drinking Water

C. County building departments

D. Private labs

Answer: B

Treatment plant operators in California must:

A. Have no certification

B. Be certified at or above the grade of the facility

C. Only pass a driving test

D. Be certified only for wastewater

Answer: B

Daily operational records, such as flows, residuals, and turbidity, are important because they:

A. Are not needed once SCADA is installed

B. Document compliance and help in troubleshooting and audits

C. Only help billing

D. Replace lab testing

Answer: B

A public water system under SDWA is generally defined as a system that:

A. Serves bottled water only

B. Provides water for human consumption to at least a minimum number of people or service connections

C. Treats only wastewater

D. Provides irrigation water only

Answer: B

Which of the following is most likely regulated with an MCL rather than a treatment technique?

A. Pathogen removal

B. Lead at the tap

C. Nitrate concentration in finished water

D. Corrosion control

Answer: C

When new contaminants are considered for regulation, EPA must consider:

A. Only operator preference

B. Health risks, occurrence, and whether regulation provides meaningful risk reduction

C. Only lab capacity

D. Only taste and odor

Answer: B

Under SDWA, systems must provide Consumer Confidence Reports, or CCRs, to:

A. Internal staff only

B. Customers, summarizing water quality and any violations

C. Only regulators

D. Only at plant startup

Answer: B

California may set its own MCLs for emerging contaminants such as PFAS, which can be:

A. Less strict than EPA levels only

B. Equal to or more stringent than federal standards

C. Ignored by operators

D. For wastewater only

Answer: B

If a system consistently exceeds an MCL for a contaminant, the operator should:

A. Ignore it unless taste changes

B. Report as required, evaluate treatment changes, and follow corrective actions

C. Turn off SCADA

D. Only increase pH

Answer: B

The main purpose of setting MCLGs is to:

A. Provide a technical basis for MCLs using health-based criteria

B. Replace regulations

C. Set aesthetic standards

D. Control energy costs

Answer: A

Action levels for lead and copper are different from MCLs because they:

A. Are non-enforceable

B. Trigger specific actions, such as corrosion control and public education, instead of immediate MCL violations

C. Apply only to raw water

D. Replace all testing

Answer: B

Under the Surface Water Treatment Rule, failure to meet turbidity performance requirements is considered:

A. A secondary standard issue only

B. A treatment technique violation

C. Unrelated to SDWA

D. Only aesthetic

Answer: B

Maintaining adequate disinfectant residual throughout the distribution system is required to:

A. Control hardness

B. Control microbial contamination and meet treatment technique requirements

C. Improve taste only

D. Reduce turbidity

Answer: B

An example of a secondary contaminant is:

A. Giardia

B. Nitrate

C. Iron causing staining

D. Lead

Answer: C

Operator certification programs, like California's, are required under SDWA to ensure:

A. Only managers are trained

B. Water systems have qualified personnel to run facilities safely and in compliance

C. Only lab staff are qualified

D. Only SCADA vendors are certified

Answer: B

The best way for a T2 operator to stay current on regulations is to:

A. Ignore regulatory updates

B. Attend training, read regulatory notices, and review state guidance regularly

C. Rely only on coworkers' opinions

D. Ask customers

Answer: B

CHAPTER 10

WATER MATH

Water Operator Certification

Water math is an important part of operator certification. For T2, operators should know common units, basic conversions, dosage formulas, detention time, dilution, pressure/head conversion, and pump horsepower formulas.

Common Units and Abbreviations

- Square feet = ft^2.
- Cubic feet = ft^3.
- Cubic feet per second = ft^3/s or CFS.
- Acre-feet = ac-ft.
- Gallons per acre-foot = gal/ac-ft.
- Inches per foot = in/ft.
- Mile = mi.
- Feet per mile = ft/mi.
- Gallons per cubic foot = gal/ft^3.
- Pounds per gallon = lb/gal.
- Pounds per square inch = psi.
- Gallons per day = gpd.
- Gallons per minute = gpm.
- Million gallons = MG.
- Million gallons per day = MGD.

These are the basic water math units used throughout the chapter.

CHLORINE DOSE

A plant has a chlorine demand of 1.2 mg/L and wants a residual of 0.8 mg/L. What dose is needed?

Formula:

Dose = demand + residual Solution:

Dose = 1.2 + 0.8

Dose = 2.0 mg/L

Answer: 2.0 mg/L

POUNDS PER DAY

A water plant treats 2.5 MGD with a dose of 3 mg/L. How many pounds of chemical are fed per day?

Formula:

lb/day = MGD × mg/L × 8.34 Solution:

lb/day = 2.5 × 3 × 8.34 lb/day = 62.55

Answer: 62.55 lb/day

A system feeds 41.7 lb/day of chlorine at 2 mg/L. What is the flow in MGD?

Formula:

lb/day = MGD × mg/L × 8.34 Solution:

41.7 = MGD × 2 × 8.34

MGD = 41.7 ÷ (2 × 8.34)

MGD = 2.5

Answer: 2.5 MGD

DILUTION

You have a 10% solution and need 100 gallons of 2% solution. How many gallons of 10% solution are needed?

Formula: $C_1V_1 = C_2V_2$ Solution:

$10 \times V_1 = 2 \times 100$

$10V_1 = 200$

$V_1 = 20$

Answer: 20 gallons

PRESSURE AND HEAD

A pressure reading is 30 psi. How many feet of head is that?

Conversion:

1 psi = 2.31 ft of head Solution:

30 × 2.31 = 69.3 **Answer: 69.3 ft of head**

FLOW AND VOLUME

A pump delivers 120 gpm for 45 minutes. How many gallons are pumped?

Formula:

Gallons = gpm × minutes Solution:

Gallons = 120 × 45

Gallons = 5,400

Answer: 5,400 gallons

LIQUID CHEMICAL FEED

A liquid chemical is fed at 15 gallons per day. If the solution concentration is 4% and the specific gravity is 1.0, what is the approximate weight of chemical fed per day?

Step 1: Find pounds per gallon of solution.

Water weighs 8.34 lb/gal.

Chemical strength = 4% = 0.04

Step 2: Find pounds of chemical per gallon.

$8.34 \times 0.04 = 0.3336$ lb/gal **Step 3: Find pounds per day.**

$15 \times 0.3336 = 5.004$

Answer: about 5 lb/day

DETENTION TIME

A tank holds 3,000 gallons and flow is 150 gpm. How many minutes of detention time does it provide?

Formula:

Time = Volume ÷ Flow Solution:

Time = 3,000 ÷ 150

Time = 20

Answer: 20 minutes

DOSAGE FROM POUNDS PER DAY

A plant feeds 50 lb/day of chemical into 1.5 MGD. What is the dosage in mg/L?

Formula:

lb/day = MGD × mg/L × 8.34 Solution:

50 = 1.5 × mg/L × 8.34 50 = 12.51 × mg/L mg/L = 50 ÷ 12.51 mg/L = 3.999 **Answer: 4.0 mg/L**

PRODUCT PURITY

A plant needs 25 lb/day of active chemical. The chemical is 80% pure. How many pounds per day of product must be fed?

Formula:

Product needed = active chemical needed ÷ purity Solution:

Product needed = 25 ÷ 0.80

Product needed = 31.25

Answer: 31.25 lb/day

GPM TO GPD

A pump delivers 185 gpm. How many gallons per day is that?

Formula:

gpd = gpm × 1,440 Solution:

gpd = 185 × 1,440 gpd = 266,400

Answer: 266,400 gpd

BASIN DETENTION TIME

A basin is 30 ft long, 18 ft wide, and 12 ft deep. Flow is 1,080 gpm. What is the detention time in minutes?

Step 1: Find volume in cubic feet.

Volume = length × width × depth

Volume = 30 × 18 × 12

Volume = 6,480 ft^3

Step 2: Convert cubic feet to gallons.

Gallons = ft^3 × 7.48

Gallons = 6,480 × 7.48

Gallons = 48,470.4 gallons **Step 3: Find detention time.** Time = volume ÷ flow

Time = 48,470.4 ÷ 1,080

Time = 44.9 minutes

Answer: 44.9 minutes

DILUTION WITH STOCK SOLUTION

You need 300 gallons of 12% solution using 30% stock. How many gallons of stock are needed?

Formula: $C_1V_1 = C_2V_2$ Solution:

$30 \times V_1 = 12 \times 300$

$30V_1 = 3,600$

$V_1 = 120$

Answer: 120 gallons

PRESSURE TO HEAD

A pressure gauge reads 48 psi. What is the equivalent head in feet?

Conversion: 1 psi = 2.31 ft Solution:

48 × 2.31 = 110.88 ft

Answer: 110.9 ft

COMBINED PUMP FLOW

One pump runs at 140 gpm and a second pump runs at 95 gpm. How many gallons are pumped in 8 hours?

Step 1: Add the flows.

140 + 95 = 235 gpm

Step 2: Convert hours to minutes. 8 hours × 60 = 480 minutes **Step 3: Find total gallons.**

Gallons = 235 × 480

Gallons = 112,800

Answer: 112,800 gallons

TANK DRAWDOWN

A hypochlorite tank is 18 inches in diameter and 3 feet tall. If the tank level drops 6 inches in 2 hours, how many gallons per hour are being used?

Step 1: Convert diameter and radius to feet.

Diameter = 18 in = 1.5 ft

Radius = 0.75 ft

Step 2: Find cross-sectional area.

Area = 3.14 × r^2

Area = 3.14 × 0.75^2

Area = 1.77 ft^2

Step 3: Convert the level drop to feet.

Drop = 6 in = 0.5 ft

Step 4: Find volume used in cubic feet.

Volume = area × drop

Volume = 1.77 × 0.5

Volume = 0.885 ft^3

Step 5: Convert cubic feet to gallons.

Gallons = 0.885 × 7.48

Gallons = 6.62 gallons

Step 6: Find gallons per hour.

6.62 ÷ 2 = 3.31

Answer: about 3.3 gallons per hour

WATER HORSEPOWER

A pump moves 620 gpm against a TDH of 84 ft. What is the water horsepower?

Formula:

WHP = (Q × H) ÷ 3960

Q = flow in gpm

H = total dynamic head in feet **Solution:**

WHP = (620 × 84) ÷ 3960

WHP = 52,080 ÷ 3960

WHP = 13.15

Answer: 13.2 WHP

BRAKE HORSEPOWER

A pump delivers 450 gpm at 110 ft TDH with 78% pump efficiency. What brake horsepower is required?

Step 1: Find water horsepower.

WHP = (Q × H) ÷ 3960

WHP = (450 × 110) ÷ 3960

WHP = 12.5

Step 2: Divide by pump efficiency.

BHP = WHP ÷ pump efficiency

BHP = 12.5 ÷ 0.78

BHP = 16.0

Answer: 16.0 BHP

MOTOR HORSEPOWER

A pump delivers 500 gpm at 95 ft TDH. Pump efficiency is 82% and motor efficiency is 90%. What motor horsepower is needed?

Formula:

Motor HP = (Q × H) ÷ (3960 × pump efficiency × motor efficiency) Solution:

Motor HP = (500 × 95) ÷ (3960 × 0.82 × 0.90)

Motor HP = 47,500 ÷ 2,922.48

Motor HP = 16.3

Answer: 16.3 motor HP

FINDING FLOW FROM WATER HORSEPOWER

A pump develops 18 WHP at 72 ft TDH. What flow rate is it moving?

Formula:

WHP = (Q × H) ÷ 3960 Solution:

18 = (Q × 72) ÷ 3960

Q = (18 × 3960) ÷ 72

Q = 990

Answer: 990 gpm

TOTAL DYNAMIC HEAD

A pump suction and discharge system has 42 ft static head, 18 ft friction loss, and 6 ft extra head loss. What is TDH?

Formula:

TDH = static head + friction loss + extra head loss Solution:

TDH = 42 + 18 + 6

TDH = 66

Answer: 66 ft TDH

POWER COST

A pump requires 12.5 HP and runs 10 hours per day. If the motor is 90% efficient and electricity costs $0.14 per kWh, what is the daily power cost?

Conversion:

1 HP = 0.746 kW

Step 1: Convert output HP to input HP.

Input HP = 12.5 ÷ 0.90

Input HP = 13.89 HP **Step 2: Convert HP to kW.**

kW = 13.89 × 0.746 kW = 10.36

Step 3: Find daily energy use.

kWh = 10.36 × 10 kWh = 103.6

Step 4: Find daily cost.

Cost = 103.6 × 0.14

Cost = 14.50

Answer: $14.50 per day

CHEMICAL FEED WITH PURITY

A plant treats 5.2 MGD and needs a 3.5 mg/L dose of active chlorine. The chemical is 65% HTH, or high-test hypochlorite. How many pounds of HTH product per day are required?

Step 1: Find pounds per day of active chlorine.

lb/day = MGD × mg/L × 8.34 lb/day = 5.2 × 3.5 × 8.34 lb/day = 152.29

Step 2: Divide by purity.

Product needed = 152.29 ÷ 0.65

Product needed = 234.3

Answer: 234.3 lb/day of HTH product

CLARIFIER DETENTION TIME

A water plant has a flow of 4.0 MGD. Half goes into each of two clarifiers. Each clarifier is 60 ft in diameter and 12 ft deep. What is the detention time in hours for each clarifier?

Step 1: Find the flow to each clarifier.

4.0 MGD ÷ 2 = 2.0 MGD

Step 2: Find clarifier volume in cubic feet.

Volume = 0.785 × D^2 × depth

Volume = 0.785 × 60 × 60 × 12

Volume = 33,912 ft^3

Step 3: Convert cubic feet to gallons.

Gallons = 33,912 × 7.48

Gallons = 253,662 gallons **Step 4: Convert gallons to MG.**

253,662 gallons = 0.254 MG

Step 5: Find detention time in days.

Time = volume ÷ flow

Time = 0.254 ÷ 2.0

Time = 0.127 days

Step 6: Convert days to hours.

0.127 × 24 = 3.05 hours

Answer: 3.05 hours

WIRE-TO-WATER EFFICIENCY

A pump is 78% efficient and the motor is 92% efficient. What is the wire-to-water efficiency?

Formula:

Wire-to-water efficiency = pump efficiency × motor efficiency Solution:

0.78 × 0.92 = 0.7176 Convert to percent:

0.7176 × 100 = 71.76%

Answer: 71.8%

CHAPTER 10 REVIEW

TRUE / FALSE

1. The formula for pounds per day is:

MGD × mg/L × 8.34

ANSWER: **TRUE**

2. 1 psi equals 2.31 feet of water head.

ANSWER: **TRUE**

3. The water horsepower formula is:

WHP = (Q × H) ÷ 3960

ANSWER: **TRUE**

4. Brake horsepower equals WHP × pump efficiency.

ANSWER: FALSE

CORRECT FORMULA: BHP = WHP ÷ PUMP EFFICIENCY

5. Motor horsepower is calculated as:

Motor HP = (Q × H) ÷ (3960 × pump efficiency × motor efficiency)

ANSWER: **TRUE**

6. Detention time equals volume divided by flow.

ANSWER: **TRUE**

7. 1 MGD equals 694.4 gpm.

ANSWER: **TRUE**

8. Cubic feet to gallons conversion uses 7.48 gal/ft^3.

ANSWER: **TRUE**

9. The circular tank volume formula is:

0.7854 × D^2 × depth

ANSWER: **TRUE**

10. Pump efficiency equals BHP ÷ WHP.

ANSWER: **FALSE**

CORRECT FORMULA: PUMP EFFICIENCY = WHP ÷ BHP

11. Wire-to-water efficiency equals pump efficiency × motor efficiency.

ANSWER: **TRUE**

12. 1 horsepower equals 0.746 kW.

ANSWER: **TRUE**

13. Flow in gpm equals gallons divided by minutes.

ANSWER: **TRUE**

14. Total Dynamic Head equals static head plus friction loss.

ANSWER: **TRUE**

15. The dilution formula is:

$C_1V_2 = C_2V_1$

ANSWER: **FALSE**

CORRECT FORMULA: $C_1V_1 = C_2V_2$

16. Filter loading rate equals flow divided by filter area.

ANSWER: **TRUE**

17. Chlorine dose equals demand plus residual. Answer:

ANSWER: **TRUE**

18. Residual equals dose minus demand.

ANSWER: **TRUE**

19. 8.34 is the weight of water in pounds per gallon.

ANSWER: **TRUE**

20. 1 day equals 1,440 minutes.

ANSWER: **TRUE**

21. 1 hour equals 3,600 seconds.

ANSWER: **TRUE**

22. The specific gravity of water is 1.0.

ANSWER: **TRUE**

23. If a pump delivers 1,000 gallons in 20 minutes, the flow rate is 50 gpm.

ANSWER: **TRUE**

1,000 ÷ 20 = 50 gpm

24. A tank with 50,000 gallons and 1,000 gpm flow has 40-minute detention time.

ANSWER: **FALSE**

Correct calculation: 50,000 ÷ 1,000 = 50 minutes

25. 2.5 MGD × 3.0 mg/L × 8.34 = 62.55 lb/day.

ANSWER: **TRUE**

26. If 58.8 lb/day is fed into 3.0 MGD, the dosage is 2.35 mg/L.

ANSWER: **TRUE**

27. A 60-ft diameter tank that is 12 ft deep has a volume of about 33,930 ft^3.

ANSWER: TRUE

$0.7854 \times 60^2 \times 12 = 33{,}929.28\ FT^3$

28. 720 gpm × 95 ft head ÷ 3960 = 17.3 WHP.

ANSWER: **TRUE**

29. 600 gpm × 120 ft ÷ (3960 × 0.75) = 24.2 BHP.

ANSWER: **TRUE**

30. 14.8 WHP ÷ 20.5 BHP = 72.2% pump efficiency.

ANSWER: **TRUE**

31. 0.78 × 0.92 = 71.8% wire-to-water efficiency.

ANSWER: **TRUE**

32. 18 WHP ÷ 0.65 = 27.7 BHP.

ANSWER: **TRUE**

33. 4.2 MGD × 2.5 mg/L × 8.34 = 87.57 lb/day.

ANSWER: **TRUE**

34. 80,000 gallons ÷ 763.8 gpm = 104.7 minutes detention time.

ANSWER: **TRUE**

35. 30 ft × 18 ft = 540 ft^2 filter area.

ANSWER: **TRUE**

36. 200 mg/L × 0.0476 MG × 8.34 = 79.5 lb of chlorine.

ANSWER: **TRUE**

37. 5 cfs = 2,244 gpm.

ANSWER: **TRUE**

5 × 448.8 = 2,244 GPM

38. 7.2 cfs × 448.8 = 3,231 gpm.

ANSWER: **TRUE**

39. 12 hours × 60 minutes = 720 minutes.

ANSWER: **TRUE**

40. 1,428,100 gallons ÷ 720 minutes = 1,984 gpm.

ANSWER: **TRUE**

41. If purity is 65%, divide the active chemical required by 0.65 to find product weight. ANSWER: TRUE

42. Mixed-flow alkalinity uses this formula:

$(C_1Q_1 + C_2Q_2) \div Q_{total}$

ANSWER: **TRUE**

43. 100 ft of head at 800 gpm requires less horsepower than 150 ft of head at the same flow.

ANSWER: **TRUE**

44. Pump efficiency can exceed 100%.

ANSWER: **FALSE**

45. Motor efficiency typically ranges from 80% to 95%.

ANSWER: **TRUE**

46. Typical pump efficiency ranges from 50% to 85%.

ANSWER: **TRUE**

47. Chemical cost equals: lb/day × days × price per lb

ANSWER: **TRUE**

48. WHP = (Q × H) ÷ 3960.

ANSWER: **TRUE**

49. BHP = WHP × pump efficiency.

ANSWER: **FALSE**

Correct formula: **BHP = WHP ÷ pump efficiency**

50. Motor HP = (Q × H) ÷ (3960 × pump efficiency × motor efficiency).

ANSWER: **TRUE**

51. Wire-to-water efficiency equals pump efficiency plus motor efficiency.

ANSWER: **FALSE**

Correct formula: **Wire-to-water efficiency = pump efficiency** × **motor efficiency**

52. TDH = static head + friction loss.

ANSWER: **TRUE**

53. 1 cfs = 448.8 gpm.

ANSWER: **TRUE**

54. Higher flow requires more horsepower.

ANSWER: **TRUE**

55. Higher head requires more horsepower.

ANSWER: **TRUE**

56. 1 HP = 0.746 kW.

ANSWER: **TRUE**

57. Daily energy cost equals kW × hours × price per kWh.

ANSWER: **TRUE**

58. Pump curves show the relationship between head and flow.

ANSWER: **TRUE**

59. VFDs, or variable frequency drives, can save energy by adjusting pump speed.

ANSWER: **TRUE**

FINAL EXAM PREPARATION

MOST COMMON EXAM WATER TREATMENT QUESTIONS AND ANSWERS

THIS SECTION IS FOR FULL-BOOK EXAM PREPARATION AND COVERS TASTE AND ODOR, WATER MATH, COAGULATION, SEDIMENTATION, FILTRATION, DISINFECTION, PUMPS, AND GENERAL TREATMENT CONCEPTS.

What is the #1 cause of taste and odor complaints in water distribution systems?

A. Algae growth

B. Cross connections

C. Corroded pipes

D. High chlorine residual

Answer: B

Which oxidant is least effective for taste and odor control?

A. Chlorine

B. Potassium permanganate

C. Chlorine dioxide

D. Hydrogen peroxide

Answer: B

What is the maximum acceptable Threshold Odor Number, or TON, for drinking water?

A. 1

B. 2

C. 3

D. 5

Answer: C

What odor describes hydrogen sulfide, or H_2S, in water?

A. Musty

B. Chlorinous

C. Rotten egg

D. Metallic

Answer: C

When investigating taste and odor complaint, what is the first step?

A. Test chlorine residual at a hydrant

B. Interview with the customer

C. Flush the distribution main

D. Check the tank

Answer: B

Which descriptor indicates algae growth in source water?

A. Rotten egg

B. Metallic

C. Musty or earthy

D. Chemical

Answer: C

What is the minimum chlorine residual required to help prevent biological regrowth?

A. 0.1 mg/L

B. 0.5 mg/L

C. 1.0 mg/L

D. 2.0 mg/L

Answer: B

Which treatment method is best for removing hydrogen sulfide?

A. Activated carbon

B. Aeration

C. Potassium permanganate

D. Chlorine dioxide

Answer: B

When multiple customers report the same odor, what should the operator do first?

A. Assume it is a plumbing issue

B. Test customer water heaters

C. Determine if it is a distribution system or source issue

D. Flush all mains immediately

Answer: C

Which parameter helps detect biological regrowth in distribution pipes?

A. Turbidity

B. pH

C. Chlorine residual

D. Temperature

Answer: C

Algae odors are most common during which season?

A. Winter

B. Spring

C. Summer or fall

Answer: C

What usually describes metallic taste in water?

A. Algae

B. Corroded plumbing pipes

C. High chlorine

D. Low pH

Answer: B

Which odor indicates chlorine residual is too high?

A. Musty

B. Chlorinous

C. Rotten egg

D. Fishy

Answer: B

What is the best way to prevent cross connections?

A. Increase chlorine

B. Install backflow preventers

C. Flush mains weekly

D. Clean tanks monthly

Answer: B

Working upstream means checking:

A. Main → tank → source → customer

B. Customer → service line → main → tank → source

C. Source → tank → main → service line → customer

D. Tank → main → customer → service line

Answer: B

Which odor is associated with decaying organic matter?

A. Metallic

B. Muddy or earthy

C. Chlorinous

D. Rotten egg

Answer: B

Storage tanks should usually be cleaned every:

A. 1–2 years

B. 3–5 years

C. 6–8 years

D. 10 years

Answer: B

Which chemical is an effective algaecide for surface water?

A. Alum

B. Copper sulfate

C. Soda ash

D. Lime

Answer: B

What causes fishy odor in water?

A. Hydrogen sulfide

B. Specific algae types

C. Corroded pipes

D. High chlorine

Answer: B

Green or oily odor typically indicates:

A. Metallic pipes

B. Algae in surface water

C. Chemical contamination

D. Low pH

Answer: B

What is the best treatment for organic chemical odors?

Activated carbon

Potassium permanganate

UV

Answer: A

Chemical odor from a single customer is usually caused by:

A. Distribution system

B. Source water

C. Customer-side plumbing

D. Storage tank

Answer: C

What is the formula for pounds per day?

A. MGD + mg/L + 8.34

B. MGD × mg/L × 8.34

C. MGD ÷ mg/L ÷ 8.34

D. mg/L × 8.34 ÷ MGD

Answer: B

1 psi equals how many feet of water head?

A. 1.0

B. 2.31

C. 3.45

D. 5.0

Answer: B

What is the water horsepower formula?

A. (Q × H) ÷ 3960

B. (Q × H) × 3960

C. Q ÷ (H × 3960)

D. H ÷ (Q × 3960)

Answer: A

What is the brake horsepower formula?

A. WHP × pump efficiency

B. WHP ÷ pump efficiency

C. WHP + pump efficiency

D. PUMP EFF

Answer: B

What is the detention time formula?

A. Flow ÷ volume

B. Volume ÷ flow

C. Volume × flow

D. Volume + flow

Answer: B

1 MGD equals how many gpm?

A. 1,000

B. 694.4

C. 448.8

D. 1,440

Answer: B

Cubic feet to gallons uses which conversion?

A. 7.0 gal/ft^3

B. 7.48 gal/ft^3

C. 8.34 gal/ft^3

D. 6.28 gal/ft^3

Answer: B

What is the circular tank volume formula?

A. $\pi \times D \times$ depth

B. $0.7854 \times D^2 \times$ depth

C. $D^2 \times$ depth

D. $\pi \times D^2 \times$ depth

Answer: B

1 horsepower equals how many kW?

A. 0.546

B. 0.746

C. 1.0

D. 1.5

Answer: B

Flow in gpm equals:

A. Gallons ÷ minutes

B. Gallons × minutes

C. Gallons + minutes

D. Gallons – minutes

Answer: A

Total Dynamic Head, or TDH, equals:

A. Static head - friction loss

B. Static head + friction loss

C. Static head × friction loss

D. Friction loss ÷ static head

Answer: B

What is the dilution formula?

A. $C_1V_2 = C_2V_1$

B. $C_1V_1 = C_2V_2$

C. $C_1 \div V_1 = C_2 \div V_2$

D. $C_1 + V_1 = C_2 + V_2$

Answer: B

Filter loading rate equals:

A. Flow ÷ filter area

B. Flow × filter area

C. Filter area ÷ flow

D. Flow + filter area

Answer: A

What is the chlorine dose formula?

A. Demand - residual

B. Demand + residual

C. Demand × residual

D. Residual ÷ demand

Answer: B

What is the residual formula?

A. Dose + demand

B. Dose - demand

C. Dose × demand

D. Dose ÷ demand

Answer: B

Minimum acceptable chlorine residual is:

A. 0.1 mg/L

B. 0.5 mg/L

C. 1.0 mg/L

D. 2.0 mg/L

Answer: B

1 hour equals how many seconds?

A. 60

B. 100

C. 360

D. 3,600

Answer: D

The specific gravity of water is:

A. 0.5

B. 1.0

C. 1.5

D. 2.0

Answer: B

If a pump delivers 1,000 gallons in 20 minutes, what is the flow rate?

A. 20 gpm

B. 30 gpm

C. 50 gpm

D. 100 gpm

Answer: C

A tank with 50,000 gallons and 1,000 gpm flow has a detention time of:

A. 25 minutes

B. 40 minutes

C. 50 minutes

D. 100 minutes

Answer: C

2.5 MGD × 3.0 mg/L × 8.34 equals:

A. 31.25 lb/day

B. 62.55 lb/day

C. 93.75 lb/day

Answer: B

If 58.8 lb/day is fed into 3.0 MGD, the dosage is:

A. 1.15 mg/L

B. 2.35 mg/L

C. 3.55 mg/L

Answer: B

A 60-ft diameter tank that is 12 ft deep has a volume of:

A. 16,965 ft^3

B. 33,930 ft^3

C. 45,000 ft^3

D. 50,000 ft^3

Answer: B

720 gpm × 95 ft head ÷ 3960 equals:

A. 12.3 WHP

B. 17.3 WHP

C. 22.3 WHP

D. 27.3 WHP

Answer: B

600 gpm × 120 ft ÷ (3960 × 0.75) equals:

A. 14.2 BHP

B. 24.2 BHP

C. 34.2 BHP

D. 44.2 BHP

Answer: B

14.8 WHP ÷ 20.5 BHP gives a pump efficiency of:

A. 52.2%

B. 62.2%

C. 72.2%

D. 82.2%

Answer: C

0.78 × 0.92 gives a wire-to-water efficiency of:

A. 61.8%

B. 71.8%

C. 81.8%

D. 91.8%

Answer: B

18 WHP ÷ 0.65 equals a BHP of:

A. 17.7

B. 27.7

C. 37.7

D. 47.7

Answer: B

Pump efficiency can exceed 100%.

A-True

B-False

Answer: B

4.2 MGD × 2.5 mg/L × 8.34 equals:

A. 47.57 lb/day

B. 87.57 lb/day

C. 127.57 lb/day

D. 167.57 lb/day

Answer: B

80,000 gallons ÷ 763.8 gpm gives a detention time of:

A. 84.7 minutes

B. 94.7 minutes

C. 104.7 minutes

D. 114.7 minutes

Answer: C

30 ft × 18 ft gives a filter area of:

A. 340 ft^2

B. 440 ft^2

C. 540 ft^2

D. 640 ft^2

Answer: C

200 mg/L × 0.0476 MG × 8.34 equals how many pounds?

A. 49.5

B. 59.5

C. 69.5

D. 79.5

Answer: D

cfs equals how many gpm?

A. 1,244

B. 1,744

C. 2,244

D. 2,744

Answer: C

7.2 cfs × 448.8 equals how many gpm?

2,231

3,231

4,231

5,231

Answer: B

12 hours × 60 minutes equals how many minutes?

A. 620

B. 720

C. 820

D. 920

Answer: B

1,428,100 gallons ÷ 720 minutes gives a flow of:

A. 1,584 gpm

B. 1,684 gpm

C. 1,784 gpm

D. 1,984 gpm

Answer: D

If purity is 65%, divide active chemical by:

A. 0.65

B. 0.75

C. 0.85

D. 0.95

Answer: A

Motor efficiency typically ranges from:

A. 50–65%

B. 65–80%

C. 80–95%

D. 95–100%

Answer: C

Typical pump efficiency ranges from:

A. 30–50%

B. 50–85%

C. 85–100%

D. 100–120%

Answer: B

Chemical cost equals lb/day × days × price per lb.

A. True

B. False

Answer: A

Coagulation adds chemicals to:

A. Remove dissolved solids

B. Neutralize particle charges

C. Increase pH

D. Add chlorine

Answer: B

Which is a common coagulant chemical?

A. Sodium hypochlorite

B. Alum

C. Copper sulfate

D. Activated carbon

Answer: B

Flocculation is:

A. Rapid mixing to form large particles

B. Slow mixing to form floc

C. Adding chlorine

D. Settling particles

Answer: B

A jar test determines:

A. Optimum coagulant dose and pH

B. Chlorine residual

C. pH only

D. Turbidity only

Answer: A

Coagulant dosage usually ______ with higher turbidity.

A. Decreases

B. Increases

C. Stays the same

D. Cannot determine

Answer: B

pH affects coagulant:

A. Color

B. Effectiveness

C. Temperature

D. Storage

Answer: B

Which is another common coagulant?

A. Sodium hypochlorite

B. Ferric sulfate

C. Copper sulfate

D. Activated carbon

Answer: B

Polymer is used as a:

A. Coagulant

B. Coagulant aid

C. Disinfectant

D. Algaecide

Answer: B

Rapid mix time for coagulation is typically:

A. 1–5 minutes

B. 15–30 minutes

C. 30–60 minutes

Answer: A

Flocculation time is typically:

A. 1–5 minutes

B. 15–30 minutes

C. 1–2 hours

D. 3–4 hours

Answer: B

Over-mixing during flocculation can:

A. Improve floc

B. Break up floc

C. Increase pH

D. Add chlorine

Answer: B

Under-mixing during coagulation prevents:

A. pH adjustment

B. Particle neutralization

C. Settling

D. Filtration

Answer: B

Coagulation removes:

A. Dissolved solids

B. Suspended particles

C. Bacteria only

D. Chlorine **Answer: B**

Optimum pH for alum is typically:

A. 4.0–5.0

B. 6.0–7.5

C. 8.0–9.0

Answer: B

Lime is added to:

A. Increase coagulation

B. Adjust pH during coagulation

C. Add chlorine

D. Remove turbidity

Answer: B

Sludge from coagulation and flocculation must be:

A. Disposed of properly

B. Reused

C. Burned

D. Ignored

Answer: A

Coagulation effectiveness ______ at very low temperatures.

A. Increases

B. Decreases

C. Stays the same

D. Cannot determine

Answer: B

TDS, or total dissolved solids, affects:

A. Coagulant dose

B. Chlorine dose

C. pH only

D. Temperature

Answer: A

Polymer helps by:

A. Neutralizing charges

B. Strengthening floc

C. Adding chlorine

D. Adjusting pH

Answer: B

Rapid mixing time should be:

A. Short

B. Long

C. The same as flocculation

D. 1 hour

Answer: A

Flocculation uses:

A. High velocity

B. Low velocity

C. No mixing

D. Chlorine

Answer: B

Coagulation is most effective at a pH of:

A. 2–4

B. 6–8

C. 9–11

D. 12–14

Answer: B

Higher flow rates ______ sedimentation efficiency.

A. Improve

B. Reduce

C. Have no effect on

D. Cannot determine

Answer: B

Short-circuiting occurs when:

A. Water follows the normal path

B. Water bypasses the normal flow path

C. Water settles properly

D. Flow is slow

Answer: B

Lamella clarifiers use:

A. Vertical plates

B. Inclined plates

C. No plates

D. Chemical only

Answer: B

Sludge should be removed from clarifiers:

A. Never

B. Regularly

C. Once per year

D. Every 5 years

Answer: B

Higher settling velocity ______ particle removal.

A. Improves

B. Reduces

C. Has no effect on

D. Cannot determine

Answer: A

Temperature affects settling.

A. True

B. False

Answer: A

Cold water ______ particle settling.

A. Improves

B. Reduces

C. Has no effect on

D. Speeds up

Answer: B

Clarifier overflow rate should be:

A. Uncontrolled

B. Controlled

C. Maximum

D. Minimum

Answer: B

Sludge concentration affects:

A. Color

B. Disposal methods

C. Temperature

D. pH only

Answer: B

Turbidity should be measured after:

A. Coagulation

B. Sedimentation

C. Pre-chlorination

D. Intake

Answer: B

Floc break-up in sedimentation ______ efficiency.

A. Improves

B. Reduces

C. Has no effect on

D. Speeds up

Answer: B

Which clarifier type is more common?

A. Vertical flow

B. Horizontal flow

C. Upward flow

D. Downward flow only

Answer: B

Tube settlers ______ settling capacity.

A. Reduce

B. Increase

C. Have no effect on

D. Eliminate

Answer: B

Sludge blanket depth should be:

A. Ignored

B. Monitored

C. Kept at maximum

D. Kept at zero

Answer: B

A sludge blanket that is too thick ______ clarifier performance.

A. Improves

B. Reduces

C. Has no effect on

D. Speeds

Answer: B

Sedimentation removes approximately ______ of suspended particles.

A. 50%

B. 70%

C. 90%

D. 100%

Answer: C

Which is not a clarifier type?

A. Lamella

B. Tube settler

C. Membrane

D. Horizontal flow

Answer: C

Detention time may be measured in:

A. Seconds

B. Minutes

C. Hours

D. Any of the above

Answer: D

Surface loading should generally be:

A. High

B. Low

C. Maximum

D. Variable

Answer: B

Settling is faster in:

A. Cold water

B. Warm water

C. The same temperature

D. Cannot determine

Answer: B

Sludge removal helps prevent:

A. Short-circuiting

B. Poor settling

C. Both A and B

Answer: C

Gravity filters use:

A. Atmospheric pressure

B. High pressure

C. Vacuum

D. Momentum

Answer: A

Typical filtration rate for dual media is:

A. 1–2 gpm/ft^2

B. 2–6 gpm/ft^2

C. 10–15 gpm/ft^2

D. 20–30 gpm/ft^2

Answer: B

Filter media should be cleaned when:

A. Head loss is low

B. Head loss is high

C. Flow is high

D. Turbidity is low

Answer: B

Backwashing:

A. Cleans with forward flow

B. Reverses flow to clean media

C. Adds chlorine

D. Increases pressure

Answer: B

Mudballs form from:

A. Good cleaning

B. Poor filter cleaning

C. High flow

D. Low turbidity

Answer: B

Which is common filter media?

A. Anthracite only

B. Sand

C. Activated carbon

Answer: B

Anthracite is:

A. Heavier than sand

B. Lighter than sand

C. The same weight as sand

D. Not used in filters **Answer: B**

Dual media filters use:

A. Sand and gravel

B. Sand and anthracite

C. Carbon only

D. Membrane only

Answer: B

Membrane filters remove:

A. Larger particles

B. Smaller particles than granular media

C. The same size particles

D. No particles

Answer: B

Turbidity after filtration should be:

A. Less than 10 NTU

B. Less than 5 NTU

C. Less than 1 NTU

D. Less than 100 NTU

Answer: C

filter effluent valve:

A. Controls backwash rate

B. Controls flow rate during a filter run

C. Adds chlorine

D. Monitors pH

Answer: B

Filter run length ______ with high-turbidity source water.

A. Increases

B. Decreases

C. Stays the same

D. Cannot determine

Answer: B

Backwash water must be:

A. Reused untreated

B. Treated or disposed of properly

C. Ignored

D. Released without control

Answer: B

Which filter has higher filtration rates?

A. Slow sand filters

B. Rapid sand filters

C. Membrane filters

D. All have the same rate

Answer: B

Slow sand filters are preferable for:

A. Uniform water quality

B. Non-uniform water quality

C. High turbidity

D. Low pH

Answer: A

Media grading prevents:

A. Better removal

B. Fine particles from clogging

C. Backwashing

D. Filtration

Answer: B

Surface wash systems:

A. Add chlorine

B. Help clean filter media

C. Increase pressure

D. Reduce flow

Answer: B

Air scour is used during:

A. Filtration

B. Backwashing

C. Coagulation

D. Sedimentation

Answer: B

Filter media should usually be replaced every:

A. 1–2 years
B. 5–7 years
C. 10+ years
D. 20 years

Answer: C

Higher head loss indicates the filter is:

A. Clean
B. Dirty
C. New
D. Unused

Answer: B

Filter performance is monitored by:

A. pH
B. Turbidity
C. Temperature
D. Color

Answer: B

Precoat filtration commonly uses:

A. Sand
B. Anthracite
C. Diatomaceous earth
D. Membrane

Answer: C

Filter effluent quality should be:

A. Worse than influent
B. Better than influent
C. The same as influent
D. Cannot determine

Answer: B

Turbidity spikes indicate:

A. Good filtration

B. Filter problems

C. Low flow

D. Clean media

Answer: B

Filter media should be:

A. Uniform in size only

B. Graded properly

C. Random size

D. Any size

Answer: B

Filtration rate is measured in:

A. GPM

B. GPM/ft^2

C. ft^3

D. Gallons

Answer: B

Backwash rate should be:

A. Minimum

B. Adequate to clean

C. Maximum

D. Zero

Answer: B

Filter media type affects:

A. Only color

B. Removal efficiency

C. Temperature

D. pH only

Answer: B

Which reduces filter run time?

A. Low turbidity
B. High turbidity
C. Low flow
D. Clean media

Answer: B

Breakpoint chlorination occurs when:

A. Chlorine is added
B. Demand is met
C. pH is high
D. Turbidity is low

Answer: B

Which is safer than chlorine gas?

A. Chlorine gas
B. Liquid chlorine, or hypochlorite
C. Ozone
D. UV

Answer: B

Which method measures chlorine residual?

A. Turbidity meter
B. DPD method
C. pH meter
D. Temperature probe

Answer: B

Pre-chlorination can cause:

A. Better taste
B. THM formation
C. Lower pH
D. Less turbidity

Answer: B

UV disinfection leaves:

A. Strong residual

B. No residual

C. Chlorine residual

D. Ozone residual

Answer: B

Which is a stronger disinfectant than chlorine?

A. UV

B. Ozone

C. Iodine

D. Bromine

Answer: B

Hypochlorite solutions:

A. Gain strength over time

B. Lose strength over time

C. Stay the same

D. Cannot determine

Answer: B

Which hypochlorite is more stable?

A. Sodium hypochlorite

B. Calcium hypochlorite

C. Both are equal

D. Neither

Answer: B

Higher pH ______ chlorine activity.

A. Improves

B. Reduces

C. Has no effect on

D. Speeds up

Answer: B

Contact time is:

A. Not important

B. Critical for disinfection

C. Optional

D. The same as flow rate

Answer: B

Turbidity ______ disinfection effectiveness.

A. Improves

B. Reduces

C. Has no effect on

D. Speeds up

Answer: B

Chloramine forms when:

A. Chlorine is added

B. Chlorine reacts with ammonia

C. pH is high

D. Turbidity is low

Answer: B

Chloramine provides:

A. Shorter residual

B. Longer-lasting residual

C. No residual

D. Pungent residual

Answer: B

Dechlorination can use:

A. Chlorine

B. Activated carbon

C. Alum

D. Polymer

Answer: B

Which chemical is used for dechlorination?

A. Sodium hydroxide

B. Sodium bisulfite

C. Sodium hypochlorite

D. Sodium chloride

Answer: B

Disinfection byproducts include:

A. Only THMs

B. THMs and HAAs

C. Only HAAs

D. Neither THMs nor HAAs

Answer: B

THMs are regulated as:

A. Secondary standards

B. Primary health standards

C. Tertiary standards

D. Not regulated

Answer: B

Contact basins must be sized for:

A. Flow rate only

B. CT requirement

C. pH

D. Temperature

Answer: B

Disinfection is usually the:

A. First treatment step

B. Final treatment step

C. Middle step

D. Optional step

Answer: B

Pathogens include:

A. Bacteria only

B. Bacteria, viruses, and protozoa

C. Viruses only

D. Protozoa only

Answer: B

Giardia is:

A. More resistant to chlorine than viruses

B. Less resistant to chlorine than viruses

C. The same resistance as viruses

D. Not resistant

Answer: A

Which disinfectant does not leave a residual?

A. Chlorine

B. Chloramine

C. Ozone

Answer: C

What is the best disinfectant for Cryptosporidium?

A. Chlorine

B. Ozone

C. Chloramine

D. Iodine

Answer: B

Optimum pH for alum coagulation is:

A. 4.5–5.7

B. 6.5–7.2

C. 8.0–9.0

D. 10.0–11.0

Answer: B

What is a good process sequence for turbidity removal?

A. Coagulation, flocculation, sedimentation, filtration

B. Filtration only

C. Sedimentation only

D. Coagulation only

Answer: A

Which can serve as a primary disinfectant and provide a secondary residual?

A. UV

B. Hypochlorite

C. Chloramines

D. Ozone

Answer: B

Typical motor efficiency is:

A. 50–65%

B. 65–80%

C. 80–95%

D. 95–100%

Answer: C

Typical pump efficiency is:

A. 30–50%

B. 50–85%

C. 85–100%

D. 100–120%

Answer: B

TDH equals:

A. Static head - friction loss

B. Static head + friction loss

C. Static head × friction loss

Answer: B

Higher flow requires:

A. Less horsepower

B. More horsepower

C. The same horsepower

D. Cannot determine

Answer: B

VFDs, or variable frequency drives:

A. Waste energy

B. Save energy

C. Have no effect

D. Increase cost

Answer: B

CLOSING MESSAGE

Water treatment is a field that matters. Every operator plays a part in protecting public health and making sure communities have safe drinking water.

If you are studying for your state water operator exam, keep going. The process may feel difficult at first, but every definition, every treatment step, and every practice question brings you closer to your goal.

A state license can open the door to a good career, a better future, and a meaningful way to serve your community. Keep learning, keep practicing, and believe that your effort can change your life.

I hope this book helps you take that next step with confidence.

JIMMY Abdalla (Jimmy)

Water chemist

California

g.a.waterexams@gmail.com

www.ingramcontent.com/pod-product-compliance
Lightning Source LLC
LaVergne TN
LVHW070521061126
851857LV00023B/2127
* 9 7 9 8 9 5 1 1 3 1 4 6 1 *